Praise for Bash Dibra's *CatSpeak*

"Those seeking to teach their cats not to scratch furniture, chew on houseplants, or indulge in other unacceptable behavior can learn how in *CatSpeak*." —*Publishers Weekly*

"Dibra, an animal behaviorist and 'celebrated dog trainer to the stars,' now turns his attention to cats, wisely noting that whereas one gives a dog a command, a cat only listens to suggestions. . . . A great basic primer." —*Booklist*

"Cats have a reputation for being independent and untrainable, a myth Dibra puts to rest by showing owners how to build a relationship with their felines based on understanding and trust, with a healthy dollop of patience, persistence, and praise." —*Library Journal*

Praise for Bash Dibra's *Dog Training by Bash*

"A great book." —*Los Angeles Times Book Review*

"A helpful, easy-to-follow, complete guide." —*Booklist*

"Great fun to read . . . well-written, accurate, informative, and entertaining." —*Library Journal*

Also by Bash Dibra

DOG TRAINING BY BASH

TEACH YOUR DOG TO BEHAVE

DOGSPEAK

Bash Dibra

with Elizabeth Randolph

Illustrations by José Dennis

CatSpeak

HOW TO COMMUNICATE WITH CATS

BY LEARNING THEIR SECRET LANGUAGE

New American Library

New American Library
Published by New American Library, a division of Penguin Group (USA) Inc.,
375 Hudson Street, New York, New York 10014, USA
Penguin Group (Canada), 10 Alcorn Avenue, Toronto,
Ontario M4V 3B2, Canada (a division of Pearson Penguin Canada Inc.)
Penguin Books Ltd., 80 Strand, London WC2R 0RL, England
Penguin Ireland, 25 St. Stephen's Green, Dublin 2,
Ireland (a division of Penguin Books Ltd.)
Penguin Group (Australia), 250 Camberwell Road, Camberwell, Victoria 3124,
Australia (a division of Pearson Australia Group Pty. Ltd.)
Penguin Books India Pvt. Ltd., 11 Community Centre,
Panchsheel Park, New Delhi - 110 017, India
Penguin Group (NZ), cnr Airborne and Rosedale Roads, Albany,
Auckland 1310, New Zealand (a division of Pearson New Zealand Ltd.)
Penguin Books (South Africa) (Pty.) Ltd., 24 Sturdee Avenue,
Rosebank, Johannesburg 2196, South Africa

Penguin Books Ltd., Registered Offices:
80 Strand, London WC2R 0RL, England

Published by New American Library, a division of Penguin Group (USA) Inc.
Previously published in a Putnam edition.

First New American Library Printing, January 2003

5 7 9 10 8 6

New American Library Trade Paperback ISBN: 0-451-20709-2

The Library of Congress has catalogued the hardcover edition of this title as follows:

Dibra, Bashkim.
CatSpeak : how to learn it, speak it, and use it to have a happy, healthy, well-mannered cat/ Bash Dibra with Elizabeth Randolph ; illustrations by José Dennis.
p. cm.
ISBN 0-399-14741-1
1. Cats—Behavior. 2. Cats—Training. 3. Human-animal communication.
I. Randolph, Elizabeth. II. Dennis, José, ill. III. Title.
SF446.5.D53 2001 2001019728
636.8'0887—dc21

Designed by Carol Malcolm Russo/Signet M Design, Inc.

Printed in the United States of America

This book is dedicated to cat owners
who appreciate the unconditional
love and joy a cat brings
to their lives.

Acknowledgments

There is no way I can express my gratitude to those whose patience and understanding supported me throughout the writing of this book: my family, my friends, my clients. Special thanks go to my sister Meruet, who took over all business matters while I was "otherwise engaged," and to my sister Hope, who is always there for me. Thanks as well to my creative editor, Stacy Creamer, for keeping the book, and me, so clearly focused.

—Bash Dibra

meow meow meow meow

Contents

CatSpeak

Introduction: My Lifelong Interest in Cats

There is a fable I like: It tells us that the devil, who knew that God created the earth and all of the creatures upon it, decided to create mice in order to destroy the earth's crops and, as a result, destroy humankind. When he saw this, God created cats to destroy the devil's mice. From that day on, cats have been true friends, companions, and helpers for people.

Why am I, Bash Dibra, well-known, celebrated "*dog* trainer to the stars," writing a book about cats? What do I know about cats, and why do I want to write about them? These are questions you may ask. After all, I've written three books about dogs and their behavior,* appeared on numerous national television shows talking about dogs and demonstrating various aspects of dog training, cosponsored many dog-focused events, such as

***Dog Training by Bash* (Dutton, 1991; Signet 1992) and *Teach Your Dog to Behave* (Dutton, 1993; Signet, 1994), both with Elizabeth Randolph, and *DogSpeak,* with Mary Ann Crenshaw (Simon & Schuster, 1999).

dog walks to benefit the American Society for the Prevention of Cruelty to Animals (ASPCA)—so why am I suddenly writing about cats?

In this introduction, I'll tell you about my lifelong interest in cats and describe my considerable experience with them. I'll also explain what I hope to convey to cat owners about their pets. Anybody who knows me is well aware that I have always been keenly interested in *all* animals, what makes them tick and how to communicate with them. When I was a youngster and my family and I traveled throughout Europe as refugees from Communist Albania, we always stayed on farms in the countryside. Prior to that, while we were still in an internment camp, I had befriended the guard dogs and learned a great deal about them and their behavior. When we stayed at a farm, I spent most of my time in the barn getting to know all of the creatures there. There were always many cats, and I was intrigued with them and how different they were from the dogs I had known. I began to understand a little bit about how to "speak" to them and encourage them to respond.

When my family and I were living in New York, we always had pet cats in our household as well as dogs. After graduating from college, I was accepted at Cornell University College of Veterinary Medicine. Unfortunately, during my first year at Cornell, my father became seriously ill and died. Because I was the oldest of four children, I needed to move back home and help support my family. So I enrolled in a city college and changed my major to animal behavior.

While I was attending school part-time, I began to go to people's homes to train their dogs. Almost every dog-owning household I worked in also had at least one cat, and I often helped owners deal with cat behavior problems while I trained their dogs. This led to many referrals to other cat owners, and many members of the Veterinary Medical Association of New York City also recommended me to their cat-owning clients who had problems with their pets.

During this time, pet-cat training was not a recognized business, and I became known primarily as a dog trainer. At the same time, I formed StarPet, an animal talent agency (www.starpet.com), and trained cats for films and television commercials in both New York and California. I worked for a number of well-known advertising agencies such as Benton & Bowles, J. Walter Thompson, and Grey Advertising. Among the print and television ads for which I trained and provided cat models were many for pet-food manufacturers such as Kal Kan, Purina, Iams, and Friskies.

One of my favorite commercials was for Tender Vittles semimoist cat food. In this commercial, an owner is carrying his cat down a supermarket aisle. When the owner and his pet are in front of a Tender Vittles display, the cat reaches out a paw and pulls a package of the food off the shelf. How did I get the cat to do that? Well, I know that cats always like to climb, so I told the actor playing the cat owner to hold the cat's paws while carrying her. When they approached the shelf of cat food, I had him release one front paw, and the cat naturally put out her free paw to climb on the shelf. It looked as if she were "choosing" the product.

While working on these commercials and advertisements, I learned a great deal about how to motivate cats to do what you want them to. It was during this period that I coined a favorite saying: "To get a *dog* to do what you want, you give it a *command*. To get a *cat* to do what you want, you give it a *suggestion*." I also learned that the "Three P's," as I call them—Patience, Persistence, and Praise—work as well with cats as they do with dogs. I also like to use Compassion, Care, and Concern (the "Three C's") when I work with cats.

I provided all of the models for several spoofs of popular magazines, including *Dogue, Catmopolitan,* and *Vanity Fur,* which used both cat and dog models. Prior to working on *Catmopolitan,* I gave a seminar called "Make Your Cat a Star" at the Cat Fanciers' Association International Cat Show in Madison Square Garden, during which I put forth the first casting call for the magazine.

Currently, I contribute articles about cat behavior to publications such as *Cat Watch,* published by Cornell

University College of Veterinary Medicine, and trade magazines such as *Pet Age*, among others. I also work with Manhattan realtors to help clients' dogs behave well in their interviews with condominium boards of directors. (In many expensive condominiums, a board of directors interviews potential owners before allowing them to purchase an apartment. If there is a dog in the family, the pet must also be interviewed. Cats are exempt from examination right now, but tomorrow may be another story. Certifying your cat as a Feline Good Citizen will be a positive step in the right direction.) For more about my current work, go to www.pawsacrossamerica.com.

With the cooperation of the Cat Fanciers' Association (CFA), the largest cat registry in the world, I have developed the Feline Good Citizen test (see the Appendix). This is a partner to the Canine Good Citizen test that I helped develop for the American Kennel Club (AKC). Now a cat—any cat, pedigreed or random-bred—can earn a Feline Good Citizen certificate. This will enable her to participate in pet therapy and to enjoy travel and stays in vacation homes and hotels, as well as many other activities previously thought of as only appropriate for dogs. Perhaps your Feline Good Citizen might even become a media star! My aim is to put cats in the mainstream, where they deserve to be.

What do I hope to accomplish in writing a book about cats? I want to help cat owners learn how to communicate better with their pets, to interpret their body language, and to understand their cats' basic senses and instincts so that cats can become more a part of the family. I would like to teach cat owners a whole new way of thinking about cats and their behavior—why they do what they do—and to learn how to modify their pets' behavior in a relaxed, enjoyable way.

There is an ancient Indian legend that believes: "In the eyes of the speechless animal is a wisdom that only the truly wise can understand." I hope, through learning CatSpeak, you can begin to understand your own cat's wisdom.

Welcome to CatSpeak!

Chapter One

Discovering CatSpeak

If your cat walks into a room, looks at you, and meows loudly, do you know what she is trying to tell you? Is she hungry, or does her litter box need changing? If she suddenly jumps up from a nap and begins to caterwaul, what is the matter? Perhaps she's suffering from a hairball, or maybe there's a strange cat outside the window.

What does it mean when she opens her mouth and lets out a very soft meow while looking unblinkingly at you and purring loudly—can you guess if she's telling you she loves you? Why is she butting her head and rubbing her face against your arm or winding her tail around your leg—do you know that she's marking you as her "property" with her scent glands? Do you realize that when she pulls at your hand with her paw, claws sheathed, she wants to be petted? If you are able to interpret these and other cat gestures and vocalizations, then you really do understand cat language. If you can't, then I will help you understand CatSpeak and enable you to create a wonderful, loving relationship with your cat.

Far from being the aloof, disinterested creatures of myth, domestic cats are usually affectionate pets, responsive to their owners' wishes and lifestyles. They are extremely intuitive to human moods and behaviors. But unlike dogs, which instinctively look up to their owners as pack leaders, cats are by nature solitary creatures (the exception in the wild are lions). They do not consider their owners as "leaders," but as caring companions who provide shelter, nourishment, comfortable living conditions, and affection. They are willing to conform to their owners' wishes because they want to continue to have these things, not because they feel submissive.

That is a basic difference between dog and cat ownership. A dog responds to his owner's commands because he is pack-oriented and considers his owner the leader. A cat doesn't respond to commands—she doesn't understand them. A cat needs to be shown what is expected of her, or what is not acceptable behavior, in more subtle ways. That is not to say that a sharp "No!" or a squirt from a water bottle can't be used to stop a cat from unacceptable behavior, but more about this later.

All cats, wild and domestic, large and small, are easily recognized as felines. From the shape of their heads, the set of their ears and tails, their whiskers and graceful carriage, tigers and tabbys and all the other cats in between are definitely from the same family.

FELINE SOCIETY

As I mentioned, most cats in the wild are solitary creatures, not generally gregarious with others of their own kind. Males and females come together to mate, then the female goes off to have her kittens and raise them. As soon as they are mature, young cheetahs, leopards, pumas, and the like set out on their own to live, hunt, and find their own mates.

Lions are an exception. They live in groups, or prides, mixed males and females, and the young typically stay in their home pride when they mature. The exceptions are young males, most of which go out to seek their own females and form their own prides. Sometimes brothers travel together and become joint heads of a group of females.

CATS AND HUMANS

Cats have lived with humans for many thousands of years. The most well-documented coexistence of cats and people dates back to the Egyptians from before 2000 B.C. As we know, the Egyptians not only kept cats as pets, they also worshiped them for many reasons: Their attractiveness and grace were, of course, factors, and their usefulness in controlling rodents and snakes was much prized. Two of the distinctive feline qualities that amazed the Egyptians were the ability to see in the dark and the fact that cats' eyes change appearance noticeably in different lights. (We will talk more about that in Chapter Two.)

There are many sculptures and bronzes of Egyptian cats, and cats that belonged to aristocratic families were mummified and given magnificent funerals and elaborate tombs. Another example of the reverence for cats occurred halfway around the world. In Norse culture, the love goddess, Freya, traveled in a chariot pulled by two cats, and the Vikings are said to have taken cats on their ships with them to protect their stores from rats and mice.

Scientists now believe that the ancestors of today's domestic cats were probably the small wildcats of Africa, although European and North American wildcats may also have contributed their genes. Because cats interbreed easily, determining the exact ancestry of today's domestic cats is difficult. Cats have been bred to develop particular colors and body conformations for a mere hundred years (dogs, for centuries).

As soon as humans changed from a nomadic lifestyle and settled down to farm and live in stable locations, cats became part of almost all households. Traveling tradesmen brought cats to even the most remote parts of the world. Although they do not guard, herd, or hunt with humans, cats perform a very important and necessary service: They control the vermin population.

Early on, they protected stored grain; in towns and later cities, they also kept rodents away from food in shops and homes. In the eighteenth century, the plague, carried by rats, reared its ugly head in Europe. Cats then became particularly valued for their rodent-hunting abilities.

Their soft, affectionate natures and charming, agile appearance soon made cats household favorites. Along with dogs, cats became well-loved pets; their rodent-controlling job made them important, but their companionship endeared them to many.

Cats have many qualities that recommend them to humans. They are decorative and scrupulously clean, washing themselves thoroughly at least once a day. They are quiet, neat, and rarely house-soil. Because they are not pack-oriented, they don't require constant reassurance and rarely suffer from "separation anxiety," as dogs often do when left alone (although there are exceptions; I'll talk about one case in Chapter Nine). In short, they are ideal pets for working people, apartment dwellers, single individuals, and families with children.

SEVEN INSTINCTIVE FELINE BEHAVIORS

There are seven instinctive behaviors common to all members of the feline family. If you understand these behaviors, you will be well on the way to understanding your cat.

Flight Behavior

Cats have a highly developed sense of peripheral vision and an amazing sense of smell. If something or someone unknown—or with an unknown odor—nears, a cat will instinctively flee. Their sense of hearing is extremely acute also; a strange or unaccustomed loud noise will also trigger flight behavior. (More about cat senses in the following chapter.) Because they are able to jump amazingly high, their flight often leads them to high places.

Cats are also able to compress their bodies, allowing them to squeeze into very small spaces.

Sometimes flight behavior will propel a cat into an impossibly small space. The best advice for an owner of a cat that has squeezed herself into a crevice or small hole is to go away and leave her alone. She will usually be able to get herself out once she calms down. Flight behavior stands many cats in good stead—when confronted with a pack of marauding dogs, for example.

Chase Behavior

Cats are hunters by nature. Their keen senses impel them to chase down any small animal that moves fast and crosses their paths. Birds, lizards, bugs, mice, and rats are all fair game. Because they are able to pronate, or turn in, their front paws, they are extremely adept at grasping and "playing" with small prey—a practice some people find distasteful. But this is an instinctive act, not an act of cruelty.

Of course, chase behavior and the subsequent killing of rodents was one of the reasons that cats originally came to live with humans. Chase behavior may be turned into a game by cat owners who toss a small light ball (a Ping-Pong ball, for example) or a feathered object for a cat to bat at, chase, and "retrieve." (More about teaching your cat to retrieve in Chapter Ten.)

Hunting/Stalking Behavior:

Closely related to chase behavior, hunting behavior is a very ingrained instinct in felines. Of course, wild cats hunt for survival. This instinct has carried over into the natures of many domestic cats, especially if they are allowed outdoors. Cat owners are often distressed when their beloved pets bring home small prey, presenting a dead songbird or field mouse seemingly as a "present," very pleased with themselves and looking for praise. This is a difficult behavior to deal with if a cat is a determined hunter and is allowed outdoors—more about this later.

Territorial Behavior

Territorial marking behavior is used by cats to delineate their "home turf" and is especially prevalent among unneutered males, although neutered males and both spayed and unspayed females may engage in it. It is usually triggered by the presence of an unknown cat or cats, or an upsetting event such as a move to a new home or the arrival of a new person or animal in the house. Marking behavior consists of spraying urine against an upright surface, rather than squatting to urinate. Urinating in an inappropriate place, such as a person's bed, is another kind of territorial behavior.

Cats display territorial behavior when they scratch or claw surfaces other than designated scratching posts. Territorial behavior may also emerge as aggression toward the "offender"—a new cat, dog, or person. I'll talk more about how to deal with these behaviors in Chapter Nine.

Aggressive Behavior

Aggression is a form of defense and protection from harm. It is not common for cats to display aggressive behavior toward humans. In general, when faced with an unfriendly or threatening person, they prefer flight or avoidance. But cats can be aggressive to people and may bite or claw. A common type of aggression is chase, or predatory, aggression, in which a cat jumps out at a person as he walks by and sinks her teeth and claws into an ankle or leg.

Sometimes fear of something else (for example, an unknown cat outside) will cause a cat to strike out at a human—this is called redirected aggression. Some dominant cats will become aggressive and bite people who pet them even when seeming to have invited the petting. Cats can also be aggressive toward other cats or pets in a household. In Chapter Nine, I will tell you how to deal with an aggressive cat.

Social Behavior

Cats are not gregarious with their own kind. But domestic cats, perhaps out of a sense of necessity, or for other more subtle reasons, are usually social with their human keepers and with other household pets, some more than others.

Those who don't admire felines often remark, "Oh, sure, your cat rubs against you and acts affectionate because she wants to be fed." Perhaps this is so under certain conditions. But how to explain the purring cat that has just had dinner and lies by your side in bed or on the couch, nudging to be petted? Or the cat that greets her returning owner with happy chirps, even though she's been fed and cared for during the owner's absence? What about the pet cat that rubs affectionately against "her" dog when he's sleeping? Surely she doesn't expect the dog to feed her.

To be sure, cats do not *need* human companionship the way dogs do. They do not usually display separation anxiety or indulge in inappropriate behavior, even when left alone for long periods of time (although there are always exceptions). They seem to be quite content to be alone in familiar surroundings, especially if there is another pet in the household, ample food, water, and a clean litter tray.

However, they do seem to *enjoy* human companionship, and some highly domesticated individuals and breeds (notably Siamese and other Orientals) often *demand* it.

Vocal Behavior

Although cats regularly communicate with one another and their human companions through body language, they also use vocal sounds to communicate. From hissing to purring, cat vocalization can be subtle or demanding. Orientals, Siamese in particular, usually have loud, raucous voices, while other cats (typically longhairs) can have soft ones that are almost inaudible to humans.

A cat's meow to her owner may mean "It's time to eat," "I need to go out (or my litter tray needs changing)," "Come here and look," or a number of different things. Her purr usually means contentment or pleasure, while a hiss can be a warning to "Stop it!" or "Go away and leave me alone."

Among cats, vocalization is often more strident, taking on a threatening quality, especially if they are strangers or an unknown cat wanders into another's territory.

An attentive owner will usually be able to discern the different sounds and qualities of his cat's vocalizations. When he does, he will have gained one more step in learning CatSpeak. I will go into more detail about cat "talk" in Chapter Seven.

SIX SPECIAL FELINE ABILITIES AND IDIOSYNCRASIES

Cats have certain physical abilities that are unique to all felines.

First and perhaps most notable is the ability to *coil*

their bodies like a spring, so that they can jump very high from a stationary position. This is a great help when they are in a situation demanding flight.

Then there is the ability to turn their paws inward, or *pronate* them, due to special muscles in their front legs. This ability is useful not only when grasping prey, but also in enabling a cat to climb extremely well and quickly, using the front paws to hold on. This, too, is very useful in assisting flight.

Another physical trait that only domestic cats possess is the capacity to *retract their claws.* This enables them to walk on all types of surfaces without becoming stuck, and it also keeps their claws from wearing down in order to continuously grow, becoming excellent hunting and climbing tools and defensive weapons. (Claw care is discussed in Chapter Six.)

Cats' eyes have a great many more rods than other animals. This allows them to be able to *see in the dark*, a very handy quality for hunting rodents and other small creatures.

All cats also possess a highly developed *sense of smell.* I was made aware of this when we began photographing cats for *Catmopolitan.* The photographer used a roll of heavy paper as a backdrop and as a covering for surface the cat would sit or stand on. After the first cat model had had her pictures taken, I placed the second model on the paper. The cat went wild, fur standing on end, yowling and hissing! I quickly realized that she was reacting to the smell of a strange cat on the paper and from then on changed the paper after each model had finished. This ability to detect odors that humans can't is another very useful tool for ferreting out small prey. Also, most cats will refuse to use a litter tray that isn't clean, or one that has been used by a strange cat—more about litter-tray sanitation later. A highly developed sense of smell can also lead to a finicky appetite (see Chapter Six).

Self-grooming is another typical feline behavior. With special, extremely rough tongues, cats groom themselves several times a day, usually after a meal, when they wake up from a nap, and after they have been petted. Each cat grooms herself in the same way, and in the same order, curling her front paws upward, licking them, and then using them to do a thorough job washing her ears, face, and the top of her head. Every other part of her body is then groomed in order. If a cat should stop grooming herself, it is a sure sign to her owner that she is not feeling well. I will talk about how to groom your cat in Chapter Six.

These special feline abilities and habits are easily recognized by an observant owner and, along with their distinctive behaviors, make up the essence of a cat.

Once I understood the seven instinctive feline behaviors and the six special feline abilities and idiosyncrasies,

I was able to look at my own cats and the cats I worked with in a different light. I realized that cats' habits are completely different from dogs'. I saw that cats could not be forced to act a certain way, but that by using their natural instincts I could teach them to act the way I wanted them to (just as I used a cat's natural instincts of flight

Cats groom themselves all over continuously.

and hunting to get the cat "actor" to grab the Tender Vittles package in the commercial).

In *CatSpeak*, I hope to be able to help you learn how to channel your cat's natural instincts and abilities in order to teach her how to live contentedly and harmoniously in your household.

Chapter Two

CatSpeak for Beginners: A Cat's Special Senses

A cat has particularly keen sensory perceptions that color her view of the world around her and her reactions to it. Thanks to those senses, she sees, hears, smells, and feels things in a way very different from that of her owner, and she may react to stimuli in a way that is puzzling to humans.

For instance, why is she suddenly refusing to eat the canned food she ate last night? It was kept fresh in the refrigerator overnight. Or why is she not eating at all? What does it mean when she winds her tail around your legs as you are going down the stairs? When she butts her head against your arm as you are trying to write a check? Or when she walks on top of your computer keyboard, rubbing against you and purring when you are working? Why does she run and hide when the doorbell rings? What could she possibly be staring at behind a darkened window where nothing is visible?

These and many other questions sometimes make people think of cats as spooky. But in reality, it is merely cats' heightened sensibilities that cause them to react in ways that are unfathomable to humans until they understand the extent of cats' special senses.

CAT SENSES

In addition to the abilities and idiosyncrasies that I talked about in the last chapter, all cats have amazingly acute senses that enable them to smell, see, and hear many times more keenly than humans can. These senses play a key role in the way a cat acts and the way she interacts with her environment, including the people and other animals in it.

A Keen Sense of Smell

If people could smell as well as cats, the world would be a different place! There would be a number of places, things, and people we could not tolerate without altering their aromas. The smell-sensing, or olfactory, area of a cat's brain is many times more highly developed than that of humans!

Cats have an organ located in the roofs of their mouths called a vomeronasal, or Jacobson's, organ, that enables them to literally taste a smell by holding their mouths slightly ajar. (This capacity can be compared to serious wine tasting, in which the taster "chews" his wine.) A cat's resulting facial expression, resembling a grimace, is known as flehmen. Interestingly, this ability is shared with snakes, lizards, and some other reptiles, in which it is used to detect the presence of warm-blooded prey. (An observer will see the animal flicking out its forked tongue to pick up the surrounding scents.) In male cats, flehmen is primarily used to detect the scent of a female in heat.

This acute sense serves cats in many ways. One obvious benefit is that they are able to determine quickly whether or not food is fit to eat; sometimes food that seems just fine to an owner may cause a cat to literally turn up her nose! A sense of smell figures so strongly in a cat's food choice that if she has a chronic nasal allergy or an upper respiratory infection that stuffs up her nose, she may refuse to eat at all. The best recourse is to offer her very strong-smelling food such as fish. Also, food that is directly out of the refrigerator may not appeal to a cat because chilling takes away most of the food's odor. Cats prefer food at room temperature or slightly warmed (and it is also easier on their digestive tract). A very short "zap" in a microwave is the best way to warm up chilled cat food. Be careful not to overdo! Test the food with your fingertip before offering it to your pet.

All cats indulge in some type of scent marking many times a day, which relies entirely on their sense of smell. The most obvious and often offensive and troublesome form of scent marking is urine spraying, which is a way to mark territory and give a clear signal to other cats that "this is my property." Unneutered male cats indulge in this activity frequently, especially if there is a female in heat nearby, but even neutered males and female cats sometimes spray urine, especially if they feel threatened by an unknown cat or have moved into a new home and have a strong need to identify the place as their own. I will discuss ways to cope with this unpleasant instinct in Chapter Nine.

Other forms of scent marking that are very familiar to all cat owners are rubbing, butting (or bunting), and tail winding. Cats have glands in various parts of their bodies that secrete scents imperceptible to humans. These glands are primarily located in their cheeks, paw pads, and on the tops of their tails. They frequently butt their heads against or wind their tails around people, other animals, or objects they want to mark as their own with their particular scent. This behavior is a clear sign of affection and should not annoy an owner who understands his cat. Problems can occur when a cat winds herself around an owner's legs when he is walking down a staircase or moving around the kitchen while preparing food. This can be dangerous. A gentle nudge with your foot can usually get a cat to back off in these situations.

Whenever a cat grooms herself with her rough tongue, she is leaving the scent of her own saliva on her fur. That is why a cat will always groom herself after she has been handled or combed. She does this to remove the lingering human smell on her fur and replace it with her own special scent. A cat may also lick a favorite person or other pet for the same reason: to leave her scent on "her" property. We now know that it is not the cat's fur, or dander (dried skin), that activates an allergic reaction in some people; the dried saliva itself causes allergies.

Clawing, or scratching, is another form of scent marking. Although it also serves other purposes, the act of scratching leaves a distinctive scent on the scratched object from the glands in a cat's paw pads. That is why even declawed cats continue to "scratch" upright surfaces within their territories. Other cats readily recognize this scent. Appropriate scratching and climbing posts will provide a cat with the necessary clawing and marking materials so that she will not tear up the living room sofa. I will talk more about this in Chapter Nine.

A cat's sense of smell plays such an important part in her view of the world and how she operates in it that an owner must consider this a very important factor in his understanding of CatSpeak.

Remarkable Vision

A cat has better vision in the dark than any other mammal. This is partly because her eyes contain many more rods than cones, and rods are minuscule cells that respond to very dim light. Human eyes have more cones than rods and therefore are unable to clearly discern objects in dim light or darkness. In addition, a cat's oblong eye pupils can dilate very widely in order to allow more light into her eyes. In bright light, cats' pupils reduce to tiny slits to protect their eyes. This physical difference (between round human pupils and oblong cat pupils) was one of the many reasons cats were considered so special and were deified by the Egyptians and other cultures.

There is a special membrane behind a cat's retina, called a tapetum, which reflects light back to the retina after it has gone through once. This gives the cat a second chance to capture an image, which further enhances her night vision. You may see a bluish or yellowish glare in a cat's eyes when she is looking into bright light. This is a reflection from the tapetum.

Cats' eyes have a third eyelid. This is a membrane between the outer eyelid and the eye itself that protects and cleans the eyeball. Sometimes it may be visible in front of a cat's eye, especially when she first wakes up or if her eyes are irritated.

A cat's incredible vision colors her entire existence. But just as with humans and other animals, a cat's eyes are the windows to her feelings and thoughts. Look into your cat's eyes and you can often "read" what is going on in her head and understand more about CatSpeak.

Sharp Hearing

Cats' ears have the same internal parts as those of most mammals. Their hearing is also remarkably sensitive. With some exceptions (Scottish Folds and American Curls), cats' ears are quite large, upright, and pointed. They are made to capture the smallest sounds. Because cats are able to rotate each ear independently (would that we could!), cats are able to catch sounds from all around them. They hear high sounds much more keenly than humans or even dogs, which helps to locate small prey as it moves just below the ground or in a burrow.

Cats' hearing makes them particularly susceptible to very pronounced, shrill, harsh sounds. A cat may react with flight behavior (see Chapter One) to such seemingly innocuous sounds as the whistle of a teakettle or an especially loud doorbell. This, to her, is the equivalent of a nearby fire alarm or police siren.

Skin and Hair / Integumentary Systems

A cat's *skin* is very important. It is a shield to protect her internal organs. It also helps maintain her body temperature, because a cat perspires through her footpads. When it is cold, blood vessels in the skin contract to retain body heat. When it is hot, the skin's blood vessels dilate and help cool off the cat's body.

A cat's *hair,* or coat, forms a physical barrier between the outside environment and the cat's skin. In other words, it keeps a balance between the internal cat and the outside world. Longhaired cats have an inherited gene that stimulates vigorous hair growth.

There are three basic types of cat hairs: fine undercoat hairs, coarse guard hairs, and stiff vibrissae. The fine hairs form the soft undercoat. Longer, coarser guard hairs make up the outer coat. The stiff vibrissae stick out from the cat's body as whiskers, eyelashes, and sinus hairs (on the inside of a cat's front legs). These stiff hairs act like radar for a cat and allow her to judge spatial relationships and feel air currents. It is often observed that a cat will never try to squeeze into a space that is too small for her body; her whiskers and vibrissae measure an opening to ensure that she will not become stuck. Whiskers are also an important factor in a cat's body language. Their position indicates whether or not she is relaxed, angry, or anxious. I will discuss body language more in Chapter Seven.

When a cat is fearful or angry, her hair will stand on end all over her body (the Halloween Cat!). This is due to tiny muscles underneath the hair that raise the hair in an action known as *piloerection.* Hairless or almost hairless cats (Rexes, Spinxes, etc.) that have no guard hairs are the exceptions.

All cats shed all year long, and cats under stress always shed a great deal. Regular grooming will help prevent hairballs made of hair that is swallowed when a cat grooms herself. If a longhaired cat is not groomed regularly, the shed hair will form mats. (See more about shedding and grooming in Chapter Six.)

Feet and claws are also a part of a cat's integumentary system. Most cats have five toes on their front feet and four on the back. Some cats have extra toes on their front feet. This inherited characteristic is called polydactyly and has no effect on the cat (it is thought to have originated in Boston, with a cat off a boat from England). Cat footpads are thick and rough with sweat and scent glands in between. The scent glands play an important role in scratching behavior.

Special tendons and ligaments allow cats to extend and retract their claws. Claws serve several purposes: They are used for defense, for climbing, for catching prey, and for digging to bury waste. Claws grow rapidly and should be trimmed when necessary to prevent them from catching on things. An owner can extend a cat's claws for clipping by pressing gently on the footpads (see Chapter Six).

Both male and female cats have *anal glands* on either side of the rectum. Unlike dogs, cats rarely have problems with these glands, which empty out during a normal bowel movement. An owner may notice a slight discharge on the rim of the litter pan, which is easily removed with a wet tissue.

ARE THERE PHYSICAL DIFFERENCES AMONG BREEDS?

All domestic cats have the same physiology and the same senses. Differences in size, hair coat, and tail length, or even mutations such as tailessness (Manx), or bent ears (Scottish Fold and American Curl), do not change the basic domestic cats' sensibilities.

Chapter Three

Getting the Cat You *Really* Want

Cats come in many colors, sizes, and coat lengths, but they always have the same general body conformation. From a massive twenty-pound tiger-striped randombred cat to a petite, almost hairless eight-pound Rex, every cat has essentially the same body. The few minor differences are often in head and facial shape. A cat may have either a narrow elongated face, such as a Siamese or Abyssinian, with large, pointed ears; slanted almond-shaped eyes; a sleek, streamlined body; long legs; and a long, slim, tapered tail. Or she may have a brachycephalic (shortened and pushed-in), chubby-cheeked face, such as a Persian's, with wide-set, rounded ears; huge, perfectly round eyes that are slightly protuberant; a sturdy body; short legs; and a thicker tail.

Mutated breeds that differ from the norm in head or body type are the tailless or almost tailless Manxes; Scottish Folds, with forward-turned ears; American Curls, with backward-turned ears; and a new breed, Munchkins, which have greatly shortened legs. Most breed registries do not yet recognize Munchkins.

I will talk about how you can select the perfect pet cat for your household in Chapter Four, but here is a brief summary of some kinds of cats. There are too many cat breeds (thirty-five recognized pedigreed breeds registered with the Cat Fanciers' Association, the sponsor of the largest cat show held in the Western Hemisphere, and thirty-two breeds recognized by the Cat Fanciers' Federation, an organization based in the eastern and midwestern United States that originated in 1919) to cover in these pages, so with some exceptions, I will limit my discussion to those that are most popular according to breed registrations in CFA. The top ten pedigreed cat breeds are, in order of popularity: Persian; Maine coon; Siamese; Exotic (essentially, shorthaired Persians); Abyssinian; Oriental; Birman; Scottish Fold; American Shorthair; and Burmese.

RANDOM-BRED (MIXED-BREED) CATS

Before discussing pedigreed cats, let me talk about mixed-breed cats (or random-bred cats, as the "Cat Fancy" prefers to call them). They are also sometimes known as alley cats or house cats. Mixed breeds come in a large variety of colors and sizes. They are the result of nonselective, or random, breeding (i.e., when an unspayed female in heat is caught by an intact, or unneutered, tomcat). Many claim a purebred ancestor or two: Blue eye color or large ear size may attest to a Siamese forebear, or an extra-long coat may be traced back to a Persian or Himalayan parent or grandparent. Sometimes kittens from the same litter may look entirely different; this is called superfecundation, which is fairly common and occurs when more than one tom breeds a female cat and fertilizes her eggs during a twenty-four-hour period.

Gray or orange markings in a "mackerel" or tabby pattern; calico (usually white, black, and orange or brown); a solid gray, black, or white; or a combination of these colors are most common among mixed-breed cats.

Random-bred cats are unquestionably the most numerous and popular pet cats in the world. They are usually robust and even-tempered and are always available in shelters and pounds. In the spring (the primary kitten season), many kittens can be found in baskets and cartons at school and church fairs, "free to a good home." (See also Chapter Four for a discussion of feral cats.)

PEDIGREED CATS

Pedigreed cats have been selectively bred to maintain or enhance certain characteristics, both physical and temperamental. Because cats have not been selectively bred as long as dogs, they generally are free of the kinds of genetic diseases and disorders that many dog breeds suffer from. Exceptions are those mutated breeds such as Scottish Folds and American Curls (ears bent); tailless Manxes; those with extremely brachycephalic faces (usually Persians and Himalayans); and those that have been significantly altered, such as Munchkins, with their extremely short legs. Some of the most popular pedigreed cats are:

SHORTHAIRED CATS

Abyssinians

Abys, as they are affectionately called, have been ranked third, fourth, and fifth in popularity over the last twenty years. Although they were developed in England and

A typical shorthaired cat's face.

have been a recognized breed since the late eighteen-hundreds, Abyssinians probably look more like the sacred cats of Egypt than any other cat breed. "Ticked" coats (each hair contains one to three bands of contrasting color) and an absence of any tabby markings make the Aby's fur unique. Added to that, their long, sleek bodies; large tufted ears; long, tapering tails; pointed faces; almond-shaped eyes; and overall ruddy color make Abys easy to recognize.

They tend to be highly intelligent, active, vocal, demanding, and very social with people. An Aby is a distinct presence in a household—not to be ignored!

American Curls (Shorthaired and Longhaired)

American Curls are the result of the 1981 discovery of a stray in California with ears that curled, or bent, backward. This was because of a natural genetic mutation, and in 1983 Curls were presented to the Cat Fancy and first accepted for CFA registration in 1986. Coats can be either short or long, in all colors and patterns. Kittens are born with straight ears that curl and uncurl until the kittens are about four months old, at which point they set permanently into a curl. About 50 percent of the kittens in any litter will have straight ears (American Curl Straight Ears) and are still excellent pets or may be used for breeding. There are no known deformities associated with the curled ears.

Curls are active, inquisitive, and affectionate pets.

American, British, and European Shorthairs

American Shorthairs are number nine in popularity and, along with their British and European cousins, are often lumped together in a group called "domestic shorthairs." In 1966 the domestic shorthair (now known as the American Shorthair) was one of the five cat breeds recognized by the CFA.

These chunky cats with large, well-knit, short-legged bodies; large round eyes; and wide-set, slightly rounded ears come in all coat colors and are thought to have accompanied early settlers to this country. There are records of American Shorthairs arriving on these shores aboard the *Mayflower.* Originally brought to protect supplies from rats and humans from rat-borne disease, they soon established themselves as beloved household companions. But in the early 1900s, crossbreeding with longhaired cats and Siamese tended to dilute the original qualities of the American shorthair.

British shorthairs have a slightly brachycephalic heads and very chubby cheeks. They originated in Britain but have been bred in the United States for a number of years. These shorthaired cats go by "European" on the continent.

Shorthairs are hardy and quiet, with small voices. They are sweet, devoted to their owners, and very sociable. They are tolerant of children, other pets, and chaotic households. Some owners have said that they think they're dogs.

Burmese

Burmese cats have ranked tenth in popularity among purebred cats for six years. It is believed that they are hybrids developed in the early 1930s with a cat from Burma and a Siamese. They were first recognized by the CFA in 1936. They have solid-color coats in four recognized colors:

sable, champagne, blue (gray), and platinum. Burmese cats have large, round eyes and compact, muscular, surprisingly heavy bodies.

Extremely affectionate and sociable, Burmese are with people as much as possible. They are very talkative but have softer, less strident voices than do Siamese.

Exotics (Shorthaired and Longhaired)

Fourth and fifth in popularity over the past ten years, Exotics have been called "the best-kept secret of the Cat Fancy." They are a cross between a Persian and either an American Shorthair or a Burmese. Shorthaired Exotics resemble Persians in every regard except coats. That is, they have brachycephalic faces; short, rounded ears; and compact, sturdy bodies. Instead of long, hard-to-care-for fur, Exotics have short, plush, dense coats that do not tangle, mat, or need daily care.

Exotics are calm, gentle, and quiet. They are affectionate and enjoy company but are not as demanding of attention as Abys, Orientals, or Siamese. In general, male Exotics are more affectionate than females. Longhaired Exotics have coats with the same texture and length as Persians and can have both Persian and Himalayan coat colors.

Manx

Manxes look like domestic shorthairs, but they can have no tail ("rumpies"), a slight bump or rise in the back ("rumpy riser"), a short tail, or even a long tail. They move like rabbits because their hind legs are longer than their front legs. The Manx came from the Isle of Man in the Irish Sea; they are now found in Britain, Europe, and America.

A sturdy cat, the Manx may have either a short or long coat in a variety of colors and mixes (longhaired Manxes were formerly called Cymrics). Manxes are vocal and playful, with high energy and an endearing response to affectionate owners. They can jump extremely high due to their strong hindquarters, and many often act like dogs, retrieving toys and sometimes burying them. On the downside, Manxes are sometimes affected with spina bifida.

Orientals (Shorthaired and Longhaired)

Orientals have been ranked sixth in popularity by the CFA for the past four years; they were accepted in the CFA for competition in 1977. Their coats come in over three hundred different colors and patterns! Long, tapering heads and very large pointed ears, almond-shaped eyes, and long, slim bodies and tails give Orientals a very sleek appearance similar to that of Siamese. Shorthaired Orientals have been thought of as solid-color Siamese.

Highly sociable, curious, and intelligent, an Oriental will be a constant companion and ever-present "helper" in any household. They crave attention and are vocal and demanding by nature. An Oriental will become very depressed if she is ignored or not allowed to participate in household activities.

Rexes (Cornish and Devon)

Although not among the top ten breeds registered with the CFA, the two types of Rex cats are interesting and very popular. Both breeds have very short, wavy fur that looks rippled and feels like suede or cut velvet. They shed very little, which is caused by a genetic mutation. Large, batlike ears are set low on the sides of a Devon Rex's head, surrounding an elfin face, turned-up nose, and large, bright eyes. The Cornish Rex's ears are also large but are straight up on a small, pointed head with high cheekbones and a long nose.

Both are affectionate people-lovers, active, and demanding of attention.

Russian Blues

Russian Blues—or Archangel Cats, as they were once known—are distinguished by their thick, plush blue coats with silver tipping on each hair. Their wedge-shaped heads and large, round, bright green eyes, along with their unique coats, identify Russian Blues.

They are gentle, quiet, and undemanding and get along very well with all members of a household. Devoted to their people, Russian Blues can be both shy and playful.

Scottish Folds (Longhaired Folds)

This breed originated in Scotland in 1961 when a shepherd named William Ross saw a cat with folded ears on a farm and went on to develop the breed. Seventh and eighth in popularity in recent years, Scottish Folds' bodies, facial characteristics, and plush coats are similar to those of American and British Shorthairs, and their coat density and colors are the same. Their ears are folded in an unyielding way: That is, they bend forward at a sharp angle, the result of a spontaneous mutation. This can result in ear problems and has prevented the breed's recognition in Britain, but they are recognized by the CFA and CFF in the United States and in some European countries.

Their personalities are similar to those of the American and British Shorthairs. Scottish Folds are good-natured, gentle, adaptable cats—playful, calm, sweet-tempered, and sociable.

Siamese

High in popularity among shorthaired cats (third only to the longhaired Persians and Maine coons), Siamese cats, like Abys, are social, very vocal, and demanding, and most are affectionate. They come in four color combinations: seal points (pale fawn body and very dark brown markings); blue points (bluish-white body and slate-blue markings); chocolate points (pale fawn body and milk-chocolate markings); and lilac points (white body with pinkish-gray points).

Their bright blue eyes; long, pointed, wedge-shaped heads; and distinctive markings make them extremely attractive and immediately identifiable. Sleek, muscular, long-legged, and medium in size, with a long, straight nose; a long, straight, tapered tail; and extremely large ears, Siamese cats' silhouettes are distinct.

To quote more than one owner, "You do not own a Siamese. She owns you!" Faults among Siamese show cats are crossed eyes, a kinked tail, and white or partially white toes—but these "faults" do not affect a Siamese's pet qualities.

LONGHAIRED CATS

In addition to the longhaired versions of some of the shorthaired breeds, many purebred cats come only with a longhaired coat. Some of the most popular are:

Birmans

Birman cats probably originated in Burma, where they were considered sacred. In modern times, the breed has had a rather mysterious history. They were almost wiped out in France during World War II, but vigorous attempts to preserve them were eventually successful. They were recognized by the CFA in 1967 and have been among the top ten registered breeds in that organization for ten years.

Birmans are large, substantial cats with long, thick coats that do not mat, fluffy tails, and white "gloves." Their coats are light-colored, with dark points similar to Siamese and color-pointed Persians. Round blue eyes and a Roman nose make the Birman a very handsome cat.

A quiet, gentle nature and an active, playful personality make the Birman an ideal companion.

Himalayan (see Persian)

Maine Coons

For the past seven years, the Maine coon has been second among the top ten breeds registered with the CFA. Originating in New England, the Maine coon is a very big, handsome, solid cat with a long, shaggy coat and plumed tail that is usually held upright.

Loving, loyal, and intelligent, Maine coons are very companionable and well mannered but not terribly demanding. Even-tempered and gentle with children and other pets, Maine coons call in a soft, chirping voice when they want attention from their people.

Norwegian Forests

Norwegian Forest cats are not among the top ten breeds registered with the CFA, but they are handsome and popular pets. Relative newcomers to the United States, they were probably companions to the Vikings, protecting their grain against rodents. They were first recognized by the CFA in 1993.

Their soft, silky coats come in almost all colors and combinations. In the cold weather, their fur is fluffy and full, with an impressive mane around the neck. In warmer months, a Forest cat sheds her undercoat and appears much smaller. The tufts in their ears and exceptionally long, fluffy tails remain in all weather. Despite their full winter coats, Norwegian Forest cats require little, if any, grooming, except when they are shedding.

Norwegian Forest cats are affectionate, quiet, gentle, and very people-oriented.

Persians

Perhaps the best-known of longhaired cats, Persians have been number one in the CFA's breed registry for the past twenty years. They have broad heads; brachycephalic (pushed-in) faces; small, wide-set ears; and large, round eyes, giving them a "sweet" expression. Their long, flowing coats require daily grooming to prevent the formation of mats and to keep them beautiful. They come in varied

A brachycephalic cat's face.

colors and combinations. Red peke-faced Persians have more pronounced facial characteristics and may come in both solid colors and tabby patterns.

Affectionate, easygoing, and responsive, Persians make delightful pets.

Himalayans are not recognized separately by the CFA but are included as a division of the Persian breed. Himalayans are hybrids, the result of breeding Persians to Siamese. The result is a longhaired cat with Siamese markings. Nowadays the Himalayan is also recognized in nonpointed colors, which came from additional crossbreeding of Himalayans and Persians. Their large, round eyes, brachycephalic faces, and general body conformation are all the same as that of Persians.

"Himis" are quiet, calm, undemanding pets.

Ragdolls

Ragdolls are not among the top ten of CFA breeds, but they are interesting, popular pets. They earned their name because of an inherent characteristic in which they completely relax when picked up and become seemingly spineless, limp, and pliable, like a rag doll. They were developed in the 1960s.

Ragdolls with pointed (traditional Siamese) markings are accepted for showing, as are "mitted" (white feet and legs), bicolors (white legs, underbellies, chest, and face), and vans (dark markings on the mask, ears, and tails).

Their long fur does not have an undercoat, and they shed little, do not form mats, and require only moderate grooming. They are big, mellow, gentle cats that are very affectionate. Ragdolls often follow their owners around and tend to flop limply right underfoot. They are not jumpers and usually stay on the floor. Some will fetch and come when called. Ragdolls are great with children, who relish the breed's acceptance of being hauled around.

Somalis

Somalis are medium-size cats with soft, silky, longish coats in four recognized colors: blue, fawn, red, and ruddy. They look very much like foxes—with large ears, full ruffs, bushy tails, and masked faces—which makes them appear somewhat undomesticated.

Somalis are intelligent and active; at the same time, they are quiet and undemanding, although very extroverted and social. They have small voices and communicate with their owners through soft trills. Bursts of energy several times a day turn a Somali into a whirligig, jumping in the air, tossing balls or toys, and running sideways. Ever curious and dexterous, Somalis can open cupboards and drawers and turn on faucets.

Turkish Angoras and Turkish Vans

Longhaired Turkish Angoras were introduced to the United States in the 1950s. Originally they were all white, but they now come in many colors, including the Turkish Van (dark markings on the mask, ears, and tail).

They are extremely active and affectionate. They love to play, and some Vans are even said to enjoy swimming! Long bodies and silky coats make both varieties extremely attractive cats.

Others

There are many other shorthaired and longhaired pedigreed cats, but not room enough to describe them all. If you are interested, I suggest you go to a cat show to see the variety of breeds available today. Also, The CFA has a website (www.cfainc.org) on which you can see pictures and read descriptions of all the breeds recognized by the organization.

Chapter Four

How to Choose a Pet Cat

When you set out to choose a pet kitten or cat, the first thing you need to decide is whether you want a pedigreed or random-bred (mixed-breed) cat. In the preceding chapter, I provided an overview of the characteristics of random-bred cats and some of the many pedigreed cats.

Whatever kind of cat or kitten you are considering, there are some basic things to think about. It has often been observed that all kittens are cute and appealing. But if you want a satisfactory, loving pet cat that will be with you for many years, it is in your (and the cat's) best interests to make your choice with care.

SOCIALIZATION

If a pet kitten or cat is going to be relaxed, calm, and outgoing in a household, she needs to have learned to trust humans and be comfortable in the world around them. She should have been socialized or adapted, to have confidence in people and to rely on them for affection and daily physical needs (food, water, grooming, a clean litter pan, a warm sleeping place, and so forth).

If gentle, caring people have been part of her world from the beginning of her life, she will be comfortable with and responsive to humans. If nonthreatening children, adult cats, dogs, and other pets have been present in her early environment, she will be relaxed with them, too. Regular handling by various family members—hopefully gentle children as well as adults of both sexes—will add greatly to a cat's or kitten's well-rounded socialization. A well-cared-for, well-fed, well-socialized, calm mother cat (or queen) in good health is also an extremely important influence on her kittens.

It may be difficult to assess all of these influences, depending on the source of your intended pet, but here are a few ways to "temperament-test" a prospective feline family member.

TEMPERAMENT-TESTING A PROSPECTIVE KITTEN OR CAT

It is probably easier to test a young, playful kitten for temperament than it is to evaluate an older cat, because an adult may have become "set in her ways" already and be less apt to respond to some of your advances. I will go into more detail about this below.

The Mother and the Litter

One of the best ways to assess the temperament of a kitten is to observe her with her mother and littermates, if possible. If the mother is a calm, people-loving individual, it is likely that her kittens will be also. Kittens in a litter with their siblings should be active, playful, and alert (unless they've just had a big meal or a very active playtime and are ready for a nap). Usually, kittens' personalities are evident from a very early age. You'll see some that are extremely active, some that are aggressive, and others that are quiet and shy.

Affection/Relaxation Level

A sweet, relaxed, affectionate kitten or cat that responds to gentle handling and attention will be a pleasure to have around. A "scaredy-cat" that runs away whenever you approach her, or one that is totally disinterested in you or your ministrations, will not make a very satisfactory pet

and will probably become annoying to you after a while. In order to assess a kitten or cat's affection and relaxation level, do the following:

Pick up the kitten or cat and hold her loosely and gently (no animal likes to be held in an iron grip), one hand underneath her rump, the other around her body so she feels secure. Speak softly to her as you stroke her, and put your face near hers. An affectionate kitten or cat who is used to being handled by people will be relaxed, may cling to your clothing, purr, and even touch her nose to yours or rub her face against you. When she is put down, she will probably stay near and may even rub against your legs and act as if she wants more affection and attention.

An animal that is not used to being handled or doesn't like it will stiffen, struggle, back away from your face. She may claw frantically, meow loudly, and clearly want to be put down. When she is put down, she will run away and may look over her shoulder and hiss at you.

Activity Level

You probably want a reasonably active kitten or cat—you undoubtedly don't want a frenetic, hyperactive individual or a total coach potato. Bring along a Ping-Pong ball or other light object (a crumpled piece of stiff paper will do) and toss it on the floor. The average well-adjusted kitten will chase it, toss it in the air, bat it around frantically for a while, wrestle with it, and otherwise interact with it. An adult cat may or may not react right away and may require a little more encouragement—or perhaps she may not be interested at all. This is normal, depending on her age; some adult cats outgrow their kittenish behavior and become too dignified to play much.

A kitten that is overly shy will not react to a tossed object except to seem frightened of it; she may back off or even run away from it. She should be considered a potential problem. She may never adjust to normal household activities and sounds. She probably won't socialize much and might spend most of her time hiding in a safe place (for example, under a bed or behind the curtains).

A kitten that becomes somewhat hysterical and wild, continuing to play frenziedly, should also be considered a potential problem; she might become an irritating household presence, running around wildly at all hours, climbing draperies or bookshelves, knocking objects on the floor, and demanding attention at inappropriate times.

If one kitten in a litter seems to be much more aggressive than the others, you may want to avoid her, too. An overaggressive cat can become a problem in a household, especially if you have or ever expect to have another pet, babies, or small children in your home.

RANDOM-BRED OR MIXED-BREED CATS

Mixed-breed, or random-bred, cats can come from several sources: private homes, shelters, begging at your doorstep, or roaming in the street or neighborhood (strays or feral cats). For a first-time or inexperienced cat owner, the difference in these sources can be especially crucial. If you are looking for a family cat that is calm, affectionate, responsive to attention, accepting of other pets, children, household noise, and activity, a carefully chosen random-bred can be the answer. But you must be sure that the kitten or cat you choose has been socialized.

Kittens or Cats from Private Owners

Kittens, half-grown cats, and even fully grown cats are regularly offered for adoption on office and supermarket bulletin boards, at flea markets, in local newspapers, at church and school fairs, by word of mouth in neighborhoods, and through veterinarians. (In the latter case, you are one step ahead because you will know that the animal is in good physical health and has had all of the necessary immunizations before you acquire her.)

Private owners can be excellent sources for random-bred kittens, since you can frequently meet the mother cat and observe the litter together. This way you can observe degree of socialization, temperament, and so forth.

When an owner offers a half-grown or adult cat for adoption, you have to ask why. Is the cat destructive? Aggressive? Is she too noisy and demanding? Does she house-soil? Does she hate children? Or dogs? Or other cats? Or is the reason simply that the owner or another family member has discovered an allergy to cats? These are important questions to ask. Before you take on the ownership of a half-grown or adult cat that has been rejected by someone else, be sure to find out all of the details about why she has to be given up. If you decide to take the cat, try to work out an agreement to take her home on a trial basis for a couple of weeks so that the onus doesn't fall on you to relocate her again if she doesn't work out.

Shelters

There are several different types of shelters. The majority in most urban and suburban areas are funded by local governments. These were originally established to control both the feral and nonferal animal populations and to protect humans from roaming animals that might be vicious or carriers of animal-borne diseases (primarily rabies). That still remains their primary objective. They capture roaming raccoons, skunks, deer, and coyotes, as well as apparently homeless dogs and cats. Some shelters are operated strictly for this purpose and do not accept unwanted pets. Others serve a dual purpose and will take pets from owners, usually for a minimum donation.

These animal-control facilities are not designed to house pet animals in comfortable circumstances, but because so many people in our throwaway society decide that they do not want or can no longer keep pets, the shelters are often full to overflowing with "used" pets for adoption. In most instances, they keep these animals for a limited length of time, and older or less "adoptable" individuals are usually euthanized after a given period. Sometimes a pet escapes from its owner's property and is picked up by an animal-control vehicle, and there are sometimes sad stories of frantic owners who find their pets at an animal-control facility too late to save her—a very good reason to always be sure that a pet wears some form of identification at all times, especially if she is allowed to go outdoors.

Shelters that do not perform euthanasia at all are popularly called no-kill shelters. Because of space limitations, some accept only animals that they deem to be quickly adoptable (others obviously go elsewhere). Some well-funded organizations or those with a large volunteer staff are able to turn unadoptable pets into adoptable pets with veterinary help and/or behavior training. Still other shelters simply pack the animals in up to the bursting point—not a very pleasant existence for the animals! (Fortunately, there are few of these.)

The best of both types enlist the aid of volunteers who, along with staff members, help to make the animals' lives pleasanter and act as unofficial counselors to people seeking to adopt a pet.

There are also shelters run by private individuals or groups of people. So-called nonprofit, some are genuinely dedicated to the welfare of homeless pets and potential pet owners; others are simply thinly veiled organizations designed to raise money through appeals for funds to "help the animals." Again, the best of these shelters make use of local volunteers and do a very good job of helping animals.

The most reliable way to judge the quality of any shelter is to talk to the people who work or volunteer there. A brief conversation should make clear whether or not the people who run the facility really do have the best interests of the animals in mind and are willing and able to help you decide on the best pet for you. Well-run shelters usually take the time to keep records of the resident animals; if possible, they ask the previous owners questions about health and behavior history so they can provide potential adopters with this information. Funds are generally very limited in animal shelters. The facilities needn't be modern or shiny to be good, as long as they are clean and their charges appear well cared for.

Cats and kittens of any age and type abound in all shelters, especially in the spring, the primary breeding season. They are the result of careless owners of unspayed female cats. The choice is difficult for a potential adopter, because it is rare for a mother cat to arrive with an entire litter. Shelters differ dramatically in the degree of care, sanitation, and "creature comforts" they provide. Many cat diseases are highly contagious, and stress and overcrowding contribute to potential health problems. For starters, try to choose a shelter that has an ongoing relationship with a nearby veterinary practice. Some have an agreement that any animal chosen for potential adoption will be provided with a veterinary examination and the necessary initial immunizations or sterilization at a nominal charge.

The choice can be heartbreaking—so many cute kittens and charming older cats wanting homes! But remember to stick to the temperament-testing steps, and you may be able to avoid an unhappy mistake.

You may be closely questioned about your lifestyle and willingness to provide good lifetime care for your potential pet. For example: Are you away from home overnight a lot? If so, what will be your arrangements for cat care? Do you plan to allow your cat outdoors? If so, how will you monitor her activities, and will you make sure she's indoors at night? Do you have or plan to have a regular veterinarian who will perform annual checkups and immunizations or booster shots for the cat? Do you expect to keep this animal as a pet for the remainder of her life? Most shelters now insist on spaying or neutering any cat before adoption, or require that you have the operation performed at the appropriate time—a positive step in the right direction to avoid further pet overpopulation. You may be asked to sign an adoption agreement that details your owner responsibilities as far as the kitten or cat is concerned. If you find this a bit off-putting, as you are offering to give a home to an otherwise homeless animal, bear in mind that the shelter is trying its best to ensure that their charges survive happily and healthily in their new homes, and that they don't become yet another stray or bounce back into the shelter, rejected and traumatized by an unhappy experience.

Strays and Feral Cats

Let me say this at the outset: If you decide to take in a stray cat, it is *vital* for her to have a veterinary examination right away (before she comes into your home, especially if you have other cats or small children). Almost all stray and feral cats have parasites (fleas, ear mites, worms, etc.), and many may have diseases that are infectious to other cats. Most important, some may have diseases that are infectious to humans due to their contact with wild animals and other cats; rabies is the most life-threatening.

There are big differences between a stray and a feral cat. A *stray* has usually been someone's pet that has been either lost or abandoned. Strays are especially prevalent, for example, in resort areas in the fall. People adopt a kitten or cat for their children to play with in a vacation home, then simply drive away and leave the animal when it's time to go back to the city. There are individuals in many of these communities who make it their mission to care for these poor cats and try to find new homes for them, but it's not an easy job year after year.

Most strays are well socialized and make very good pets. They have already become accustomed to the ways of a normal household and, with luck, have had a positive experience with humans (until now). It's not too difficult to judge a stray's personality and affection level by spending some time with her. Of course, if she is a recent homeless animal unaccustomed to hunting, and her recent experience with humans has been limited to begging for handouts, she is probably starving and may be overly affectionate in hopes of some food. Try petting her *after* she has had something to eat.

When a former pet cat that is used to being housed in comfort and fed on a regular basis is forced to survive on her own for a long period of time, she will become more and more self-sufficient. Her survival instincts will become heightened, and her behavior around people will become more furtive and distrustful as she is shooed and shouted at while foraging in garbage cans for food. In short, she will have become *feral*. (Webster's dictionary defines *feral* as: "1. Of, relating to, or suggestive of a wild beast: savage; 2. not domesticated or cultivated: wild; having escaped from domestication and become wild.")

A cat such as this might be retaught to behave like a pet if she hasn't been in the wild for too long and if her former experiences as a pet were positive. With gentle, quiet coaxing and offerings of food and water, a former domestic cat that has lived as a feral may eventually learn to trust humans again and may possibly accept life inside as a house pet. Gaining the trust of such an animal can take a lot of time and effort on your part. It will be hard for her to unlearn her self-preserving ways, but if she does, she can become a loving, affectionate family member, although she still may retain a certain wariness.

On the other hand, a cat that was born of a feral mother is much more difficult to domesticate. Her entire life has been devoted to survival and her social structure has been based on living pretty much alone, or in a loose society of other cats, but not humans. Very young feral kittens *may* be tamed enough to live in peoples' homes, but their socialization can be a long and sometimes discouraging process that requires a lot of patience. Many of them remain aggressive toward humans, despite loving care.

Some people devote themselves to helping feral cats, especially in areas where there are large populations of them. Not only do they feed them, they also, in some instances, trap and take them to a veterinarian to be neutered, given a thorough examination and any necessary immunizations, and then release them where they were originally found, free to live the way they are accustomed to but not to breed and produce more feral kittens.

A third category of cats that do not live in homes but are neither strays nor feral are *country cats* or *barn cats*. By definition, they live in the country, usually on a farm or large estate. These animals never go into a house; they live instead in barns or outbuildings, underneath porches, or wherever they can find shelter. They hunt and catch rodents and are often given supplemental food by the resident humans. These cats tend to be calm and affectionate with people and other animals, but they are not truly domesticated and are not house cats. Most of these cats are uncomfortable and unhappy if they are suddenly relocated indoors. As with feral cats, a very young kitten may be an exception and adapt to indoor life.

PEDIGREED CATS

As I mentioned in Chapter Three, one of the best ways to get to know about pedigreed cats is to attend cat shows. A word of warning, however—every cat breeder naturally thinks that his or her breed will make the very best pet and usually wants to sell you one of his kittens, so be sure to keep your perspective and bring along a grain of salt when talking to breeders. People who own a particular breed as a pet may be more objective, although their opinions, too, may be slanted. Ask for references and talk to as many owners of the breed you are considering as you can. Find out about qualities that might affect your enjoyment of a particular breed, such as energy and affection levels, coat care, shedding, destructiveness in the house, idiosyncrasies, amount and volume of vocalization, and anything else that might concern you.

All pedigreed cats have to be registered in one of several national cat associations in order to be shown or bred for show. The largest of these worldwide is the Cat Fanciers' Association; other smaller and regional associations also exist. A breeder has to complete a number of registration papers and prove that his cat has the proper lineage of pedigreed ancestors. Once a cat is accepted in an association, she may be shown and bred according to that association's particular rules or breed standards (which vary). Each association recognizes a different number and variety of the existing cat breeds. For example, the CFA, the largest of the registries, recognizes thirty-five different breeds at this time, but several breeds accepted by other associations are not on the CFA's list. (By the way, a cat doesn't need a pedigree to participate in many cat shows—household pets or mixed breeds are often shown.)

Finding a Breeder for a Pet Cat

If you are like the majority of potential pet cat owners, you have little interest in either breeding or showing your cat, and you are not especially interested in a pedigree. You are looking for a particular body type, personality, and temperament that you feel will fit into your household and lifestyle and be best realized in the particular qualities of a pedigreed, rather than a random-bred, cat.

In order to find this special cat, you need to find a breeder who really loves his cats and who is willing to help you decide which individual animal will best suit you and be happiest in your home. The breeder is not, or should not be, looking for an owner who will perpetuate the breed; he wants to do that. He is looking for an owner who will provide a good home for the pet-quality (not show-quality) kittens or cats that make up the majority of his litters. He wants to find a potential owner who will *not* breed his pet and further add to whatever show faults the kitten may have (for example, a spot of white on a Siamese's paw; uneven markings on a British Shorthair; a kink in the tail of an Aby—none of which affect the cat's pet qualities). As a matter of fact, he will almost always insist that you agree to have your pet neutered. He may also ask you to keep your pet indoors at all times, not roaming free to breed accidentally, get into trouble, or be injured. He may have strong opinions about declawing a cat. Breeders often require that you promise to adhere to their particular requirements in order to qualify as an owner of one of their cats. If you don't sign on the dotted line, no kitten or cat for you. This is, of course, to force you to conform to standards of what that particular breeder considers the best home for one of his "babies."

If you are interested in showing your pet or breeding her for show-quality kittens, you have to know exactly what you are looking for and find an appropriate breeder. It may be hard to find a really good breeder who has the interests of all of his kittens in mind, not just those rare ones that are destined for show recognition and future breeding, but those that will thrive in a loving household. On the other hand, whenever a particular breed is especially popular or in style, ill-informed, opportunistic breeders will spring up. People who know nothing about feline genetics or breeding for temperament in order to produce good pets will simply breed and breed in order to produce more kittens to sell to eager pet owners. The results can be kittens that are not only physically defective but have not had the benefit of a healthy, relaxed mother. In these cases, the mother cats often become "kitten factories" and are not given the loving attention and medical care they deserve and need in order to produce healthy, well-adjusted offspring; nor are the kittens handled and socialized properly. These cat breeders are the feline equivalent of those who breed dogs in what are known as "puppy mills."

It may be that you have decided you want, or need, a kitten right away for one reason or another. But once you have found a breeder who has the exact kind of cat you have decided on, he probably won't have any kittens available right away—a good breeder doesn't produce kittens on demand. In selecting a purebred kitten or cat as a pet, the operative word should be "patience." You may have to wait quite awhile until the right pet is available.

What Should You Know About a Breeder?

Results: Word of mouth is one of the best ways to select a breeder. If you know a friend, acquaintance, friend of a friend, or coworker's sister-in-law who owns the breed of cat that you are interested in, ask questions. Is the pet satisfactory in all ways, and where did the person get it? A good breeder's reputation is based on satisfactory results over the years. A breeder who has been in the business for a reasonable amount of time and produced satisfactory pet kittens is probably reliable.

Size of Operation: How many cats does a breeder own at one time? Where do the breeder's kittens and cats spend most of their time, in cages or kennels? Are they free to roam the house? Who is the primary caretaker of the animals, the breeder or hired help? If the animals are caged or crated most of the time, does each kitten or cat have time to run around and interact with humans and other cats? (This is particularly important if you already own a cat or plan on getting two kittens at the same time.) In other words, is the breeding operation so large that the cats are constantly in cages and enclosures and have no opportunity for normal socialization with humans and other pets, and exposure to household activities; or does each animal have a chance to be held, petted, and accustomed to a normal home? This may be hard to ascertain. Obviously a breeder will want to paint the homiest picture of his cats' existence. If possible, arrange to visit the breeder and see for yourself. If not, at least ask to see photographs of the operation and find out how many cats and kittens inhabit the place at one time.

Number of Breeds: Most conscientious breeders concentrate on one breed, or on a couple of breeds that are related (for example, British Shorthairs and Scottish Folds, or a couple of Oriental breeds). If a breeder is too spread out and his operation includes a number of unrelated breeds of cat, it is impossible for him to *truly* know and understand the personalities, health concerns, and intimate details of his cats. He is not a true fancier of any one breed.

Knowledge of Health Concerns and Breeding: A good breeder naturally wants his kittens and cats to be in robust good health. But an inexperienced and/or unscrupulous breeder may not know, or may ignore, possible genetic (inherited) or congenital (present at birth) defects that may affect his breed of cat. Although these disorders are relatively rare in cats (unlike dogs, which have been selectively bred for much longer), they may affect a cat's health and suitability as a pet.

It is a good idea to read up on, or ask your veterinarian about, possible health problems in the breed you are considering and then talk to the breeder about what he is doing to help screen out any problematic genes. For example, if one of his breeding cats develops a breed-related problem, does he continue breeding, or does he neuter the animal and remove it from his breeding stock?

A few examples of potential genetic or congenital problems: Abyssinians, Siamese, and Somalis may suffer from retinal degeneration, which leads to blindness; Birmans, Himalayans, and Persians can develop cataracts; Somalis are prone to myasthenia gravis (a disease that causes muscle weakness, breathing difficulty, and regurgitation); and Himalayans often inherit von Willebrand's disease, a bleeding disorder. A breeder who is aware of these potential problems and works to prevent them is a good breeder.

What Are the Breeder's Policies About Returning Kittens or Cats? What Happens if a Kitten/Cat Sickens and/or Dies?
Any responsible breeder wants his animals to be in good health at the time they are purchased, and assumes they

are. But given the highly contagious nature of some feline diseases and the opportunity for these diseases to spread in a multicat environment or in a show milieu, a seemingly healthy kitten or cat can be secretly incubating a serious or life-threatening disease at the time of her adoption and may sicken or even die (this is especially true with young kittens that may not have received all of their immunizations by the time of purchase).

Even a carefully chosen, perfectly healthy kitten or cat may simply not work out in a household. Perhaps she is too noisy, too demanding, or not as affectionate and responsive to attention as expected. Maybe she and a family member or other pet cannot learn to get along peacefully. What are your options?

You must be sure that you are protected if any of these problems arise. Be sure the breeder provides you with a signed agreement that guarantees he will either replace the animal with another of your choice or reaccept the animal and refund your money if the placement does not work out to your satisfaction.

What Makes the Particular Kitten/Cat You Are Interested in a Pet-Quality Rather than Show-Quality Animal? In other words, what's wrong with her? If the breeder is a true professional, he or she will be able to answer this question with ease. Depending on the show standards of any breed, there are fine lines of distinction. The animal you are considering may have a tail that is too long or too short, ears that are too pointed or too rounded, or spots of color in the wrong places—all show faults that will in no way affect her suitability as a household pet.

Breed Rescue Organizations

For most recognized cat breeds, there are organizations that specialize in relocating individual cats that can no longer be taken care of by their previous owners. The reasons for this sad situation are various: an owner has fallen ill or died, or moved to a home or apartment where pets aren't permitted, or has experienced a change in family makeup, to name a few. These cats are usually well-socialized pets. In the rare instance when one is not, the breed rescue volunteers often work with the animal and try to solve any potential problems. If an individual cat is difficult in some way, the volunteers will take pains to relocate her in an appropriate environment with an experienced cat owner. Their only aim is to find good, permanent second homes for these cats.

To find out about a breed rescue group, ask your veterinarian, contact the Cat Fanciers' Association (1-732-528-9797; www.cfainc.org) or visit my website (www.pawsacrossamerica.com) to find the breed organization you are looking for. Adopting a cat from a breed rescue organization can be a perfect way to obtain the pet you want.

Chapter Five

The Homecoming

Now that you have finally found the kitten or cat you really want, it is time to prepare for her homecoming. To pave the way to a good, lasting relationship, the homecoming should be pleasant and as stress-free as possible for both of you. With this in mind, it is best if you can arrange to wait at least a day or two after choosing your pet before you bring her home so you will be able to get a few needed supplies and kitten- or cat-proof your home; you will also want to find a veterinarian and establish a proper spot for the litter pan.

You wouldn't bring a new baby home before you prepared a place for him to sleep, made sure you had all of the supplies you need in order to make his life comfortable and secure, and lined up a good pediatrician. You should give the same consideration to the new four-legged member of your family. Her first impressions of her new home will make a lot of difference in how she settles in. Don't forget that your new pet is likely to be nervous and wary. She has never been in your home before, and if she is a young kitten, she has never been away from her mother and littermates. You must provide the comfort, safety, and security she needs, as you will be both her mother and protector. She needs to feel secure and welcome. If she does, you will have gone a long way toward creating a strong bond between you and your new cat.

A VETERINARY VISIT

The most important first step is to establish a relationship with a veterinarian who will be looking after your cat's health, hopefully for the rest of her life. For the cat owner whose pet will be an integral part of the family, a steady and satisfying relationship with a veterinarian or veterinary practice is a must. The first stop on your way home with your new pet should be at the veterinarian's, especially if you are adopting a stray or a pet from a shelter. This is an absolute *must* if you have other pets or children at home; if you do not, you can wait a day or two before a veterinary visit, but no longer.

How to Choose a Veterinarian for Your Cat

If you do not already have a relationship with a veterinarian, choosing the right one may be somewhat confusing and daunting, especially if you live in an area with many veterinary practices. If you live in a small community, you may not have a choice; there may be only one veterinary practice within a reasonable distance. But in most urban and suburban areas, you may have numerous choices nearby and will need to do some research. How to find a qualified veterinarian in your area? My first suggestion is to ask responsible pet-owning neighbors and friends for recommendations. Think about what you expect from a veterinarian before you choose one.

Some Questions to Ask Yourself When Choosing a Veterinarian

- Is the veterinarian easy to talk to? Does he or she seem interested in you and your new cat?
- Does the veterinarian like cats?

- Does the hospital have facilities to keep an ill or injured cat? Are they well supervised?
- Is the veterinarian willing and able to consult with you on the telephone if necessary? Are there regular telephone hours?
- What arrangements are made in case of an emergency? Is a staff person available on weekends, holidays, and after hours, or must the cat be taken somewhere else if she gets sick or is hurt?
- Does the staff keep accurate records of immunizations and send reminders when these procedures are due?
- Does the practice perform nonmedical services such as grooming and boarding? (Note: This may not be important to you unless you live in a small community where these services are not available elsewhere.)

If all you want is routine care, yearly immunizations and boosters, and possibly routine surgery such as an ovariohysterectomy (spaying) or neutering, then the choice is not crucial. If you have limited resources, there are often low-cost clinics that are able to perform most routine procedures.

But if you are a cat owner whose pet will be an important part of your family, you will want a steady and satisfying relationship with your veterinarian. This relationship should be similar to the one you have with your personal doctor, pediatrician, or dentist. To develop this type of relationship, the veterinarian will require some time with both you and your pet in order to become knowledgeable and effective.

I am a firm believer in a group practice, rather than a single veterinarian, for a cat. In a group practice, more than one doctor will get to know your pet and will share her records. Even if you prefer to make regular appointments with one particular veterinarian in the group, you will still know that any one of the doctors at the hospital can take over if necessary. There are feline-only veterinary practitioners in some areas, which may suit you and your pet best, especially if she is a rare or unusual cat or requires special care.

Visit the clinic or hospital alone and make a personal, objective evaluation. The facility does not need to be fancy or supermodern, but are the premises clean? Are the personnel friendly and neat? Do they seem to be interested in their animal clients? What are their hours, and what provisions do they make for after-hours emergencies? What, if any, arrangements do they have with other veterinarians should your cat require a specialist such as a cardiologist, surgeon, ophthalmologist, or dermatologist? Also, what are their rates for various routine procedures? Do they accept credit cards or checks, or must you pay in cash? All of these questions are important and any reliable veterinary medical practice should have no difficulty answering them.

There are several professional associations that set standards for their member veterinary hospitals and veterinarians. The American Animal Hospital Association (AAHA; 1-800-252-2242 or www.healthypet.com) and your state or local American Veterinary Medical Association (AVMA; 1-800-248-2862 or www.avma.org) will be able to provide you with a list of veterinarians and veterinary hospitals in your area that meet their high standards.

What to Expect on the First Visit to the Veterinarian

The doctor will examine your cat for highly contagious parasites (for example, fleas, ear mites, or worms) and any infectious or contagious diseases. He will listen to her heart and lungs, take her temperature, weigh her, and give her a thorough going-over.

He will set up a schedule for your cat's all-important immunizations and booster shots and will also make a date for her neutering operation if she has not already had this performed. (I'll go into more detail about these procedures in Chapter Six.) Once the veterinarian gives your new pet a clean bill of health, you can safely introduce her to other pets and children in your household. All of the information about your cat will then be part of her permanent record, which will provide the doctor with a basis of comparison should the cat ever become ill. If you or the doctor should move in the future, this record should go along with the cat and will be invaluable to a new veterinarian.

In addition to learning details about the veterinary

practice you intend to use, you need to determine just what will be expected of *you* on your first veterinary visit. Practices often vary in the way they proceed. Call the receptionist or secretary ahead of time and ask these questions:

- Other than any medical records you may have, what else do you need to bring?
- Will the doctor want a stool sample?
- Must your cat be in a carrying case? (The answer, almost certainly, will be yes.)
- Will you be expected to restrain and hold your cat during her examination, or will a staff member do this?

The doctor will perform a thorough examination of the cat's entire body and will appreciate any information you can provide about the cat's appetite or activity level. Of course, if you take the cat to the doctor on your way home, you may not have a great deal of information except what the previous owner, or breeder, has provided.

A Cat Owner's Starter Kit

THERE ARE MANY THINGS YOU CAN DO TO EASE YOUR NEW kitten or cat's transition into your home. One of the simplest and most important is to have all of the equipment you will need *before* you bring your new pet home. Here are eight essentials to have on hand before you pick up your kitten or cat.

- Sturdy carrying case
- Food (Find out, if possible, what food your new cat is used to.)
- Food bowl (Note: any plate or bowl will do; combination food/water bowls are messy and not recommended.)
- Water bowl or bowls (There should be a water bowl available on each level of your home.)
- Litter box or boxes, litter, liners (Again, if your home has several floors, it is best to have a litter box on each level. If you already have a cat, be sure to provide at least one other litter box.)
- Sturdy, tip-proof scratching post(s), preferably burlap or sisal, or climbing "furniture" with scratching surfaces; in a large home, it is best to provide several scratching posts (Note: If your cat has been declawed, a scratching post and climbing furniture still provide necessary exercise.)
- Brush and comb
- Toys and catnip (To keep loose catnip fresh, store it in the refrigerator.)

A CARRYING CASE

You may never plan to take your cat on an extended trip, but a sturdy, roomy carrying case is an essential piece of equipment. Even if your cat learns to walk calmly on a leash, she still must travel in a carrying case for her own safety.

The most commonly used type of cat-carrying case is made of rigid, waterproof material (airline approved). There is wire on one side that provides air circulation and allows the cat to look around. Breathable carrier covers are available for cold-weather use or for especially nervous cats. Plastic covers should not be used because they cause the interior of the carrier to become too hot in warm weather or in heated areas. I much prefer a carrying case that can be opened at the top as well as the end. It is much easier to lift a nervous cat out of her case than it is to drag her out through the end (and to drop her in from the top rather than "stuff" her into the end of the case). Latches are better than zippers, which can snag fur and skin.

There are soft-sided carrying cases (Sherpa bags), which are also approved by airlines. They are lightweight, and many cat owners prefer them to hard-sided cases.

The temporary cardboard carrying cases often given to new owners by shelters and pet stores are terrible. They provide a completely false sense of security. A two-pound kitten can easily punch her way out in about five minutes—a lesson learned the hard way on the trip home by many a new cat owner when his ankles, legs, or shoulders are severely scratched by a terrified animal that has gotten loose in the car.

Before you go to pick up your new pet, be sure you

have a sturdy carrying case to put her in. You will use it throughout her lifetime.

SAFETY FIRST

Once you know your cat will be safe and secure on the trip home, you have to begin to think like a cat and make your home a safe place for her.

Cats are especially susceptible to *falls from windows,* with devastating results. This has been dubbed the "high-rise syndrome," but a fall from a second-story window can be equally damaging to a kitten or cat. There are several theories about why cats fall from windows so often, but whatever the cause, the cure is simple. Be sure there are secure screens, winter and summer, on any window that will *ever* be opened in your house (even for a few minutes). A bird flying by, a moth or leaf floating in the air—these are all irrestible to any cat, who will leap out at them with no thought of the hard pavement or earth below. Iron bars will do nothing to deter a determined cat, which can easily squeeze through. This is a most important safety measure for any cat owner.

Cats are intrigued with *linear objects*—string, yarn, needle and thread, tinsel, rubber bands. They love to play with them and often swallow them while playing. This can lead to severe intestinal damage, which may require abdominal surgery to remove the object, if caught in time—or death if not.

Look at your home through your cat's eyes, and put any such objects securely out of your cat's reach. (When you are doing this, remember that their reach is amazing: Many cats are able to open doors, boxes, drawers, cupboards, and closets, and they can also jump remarkably high.)

String that has been tied around meat, fish, or poultry is another hazard. So are any kinds of *bones*, which can do a great deal of damage to a cat's digestive tract. If your garbage contains any of these things, be sure it is covered or in a tightly closed bag—and don't leave the bag within reach unattended, since any cat can break open a plastic bag with ease!

Cats are not as apt as dogs to explore *poisonous substances* such as household cleansers because of their noxious smells. There are some exceptions, the most important being antifreeze, which has a taste cats like but contains a chemical that is *highly* toxic to cats; they need immediate veterinary help if they have ingested it. If there is a chance that antifreeze has leaked onto the floor of an attached garage, be sure to keep your indoor cat away. There is now safer nontoxic antifreeze on the market that you may want to consider.

Rat or mouse poison can be a problem if a cat catches and bites or eats a poisoned rodent. If you live in an area where these pests are a problem, and somebody may have put out poison, be especially vigilant about preventing these pests from entering your home.

A great deal has been written about potential poisoning from houseplants, but in general a cat will not eat houseplants, due to their unpleasant taste and smell. Exceptions may be when a cat, or especially a kitten, plays with fallen berries and leaves: Holly and mistletoe berries, for example, can make a cat's mouth and throat swell severely if she eats enough of them. Other potentially poisonous plants include dieffenbachia, lilies, and Japanese yew. You will not know if a new cat is prone to chewing plants, so it is best to avoid these varieties. (See Chapter Nine for ways to prevent a cat from chewing plants.)

If you think or know your cat has eaten anything poisonous or potentially poisonous, call the ASPCA National Animal Poison Control Center at 1-888-ANI-HELP (1-888-426-4435); the $30 consultation fee can be paid by credit card.

Kittens and cats usually don't chew *electrical wires,* as puppies often do. It is still a good idea to police the area where your new cat will spend time when you are away from home. Even if she does not chew wires, she may very likely play with any that are dangline invitingly, and bring a lamp, radio, or other appliance down onto her head (to say nothing of breaking the object). Unplug small appliances and tuck away their cords before you leave your new kitten or cat alone in a room with them.

THE LITTER BOX

A good, sturdy, lined litter box is just as important as a carrying case; so is its location and the litter in it. Once you have placed a litter pan in a given location and shown the cat where it is, do not move it without showing her its new location, or you may create house-soiling problems. Some behaviorists recommend moving a litter pan very gradually—only a few inches a day, so that the cat becomes used to its new placement. Most cats prefer a relatively private spot for a litter pan, and none will appreciate a busy, well-traveled area.

Many feline behaviorists recommend that the ratio of litter boxes to cats in a household should be one for each, plus one. That is, if you have two cats, you should have three litter boxes, and so forth. I feel that a great deal depends on the size of your home. If you are in a small apartment, say, and have one cat, then one easily accessible litter box is sufficient, as long as it is kept scrupulously clean.

One of the reasons for multiple litter boxes is that most cats will not use a smelly or soiled box. I recommend changing all of the litter at least once a week and removing solids whenever they appear, usually once or twice a day.

You can buy plastic liners to fit most boxes. The best are heavy-duty drawstring, for easy changing. Thin plastic can be easily torn when your cat makes her heroic efforts to bury her waste. Newspaper is not a suitable liner for a litter box. It becomes saturated and smelly and is very difficult to remove when changing the litter.

There are several types of litter available. (Again, shredded newspaper is not at all satisfactory.) Most owners prefer clay litter because a cat is less apt to scatter it. But many cats really like the sandlike clumping litter. It is very messy, and unless you have an area such as a little-used basement to put the litter tray in, you will be unhappy with the tracked sand. Even all of the pads and mats designed to clean a cat's paws do not really do the job.

Although some owners opt for a covered litter pan in the interests of appearance, most cats do not like them. Cats feel vulnerable when they are using a litter tray and prefer to be able to look around.

For soiling and litter-box-related problems, see Chapter Nine.

A SCRATCHING POST

This is another important piece of equipment for all cats. Even if your cat is declawed, she needs the stretch-

A large rope-covered scratching post.

ing exercise that she can get from using a good, sturdy scratching post or cat furniture with a scratching surface. Declawed cats regularly paw scratching posts (or your furniture, if there is no post available).

Scratching, or clawing, is an instinctive cat behavior. Cats claw for exercise, to remove loose layers of claw, and to scent-mark, as I mentioned in Chapters One and Two.

Declawing

Declawing is not generally recommended. It is a very intrusive surgery, not to be undertaken lightly. Both a veterinarian and an animal trainer should be consulted to help make a decision about whether this procedure is necessary.

In a declawing operation, a cat's claws and the toe bones to which the claws are attached are removed up to the first joint. The procedure is usually done only on the front claws, because the rear ones are rarely used for scratching (the exceptions might be for large, exotic cats). Cats are usually hospitalized for several days for observation after the operation.

After a cat goes home, her feet may remain tender for several weeks. Complications are rare, but infection may occur, causing swelling and/or a discharge from the toes, lameness, and appetite loss. Some veterinarians routinely prescribe antibiotics after the operation. It is commonly recommended that shredded paper replace litter to avoid contamination and infection.

Many reputable and concerned breeders insist that potential owners sign an agreement *not* to declaw their cats.

Follow the steps on pages 178–82 to teach your cat to use a scratching post, and you won't need to even consider declawing your pet.

A good scratching post must be sturdy enough not to tip over, no matter how vigorously a cat attacks it—and many attack their scratching posts very vigorously! That means it has to have a wide, heavy base. Plastic scratching posts are not very satisfactory, no matter what they are covered with, because they tend to tip over onto the cat. If this happens one too many times, the cat will give up and use the sofa instead. Floor-to-ceiling poles with scratching areas on them are very satisfactory.

I feel the best surfaces for a scratching post are rope and burlap, which really provide a good, resistant base for a cat to scratch. Carpet-covered posts, although they may fit best into your decor, don't work nearly as well. What's more, they accustom a cat to scratching carpet—not a habit you want to encourage.

To encourage a kitten or cat to use a scratching post, rub catnip or spray catnip scent onto its surface. I'll have more about training a cat to use a scratching post instead of your furniture in Chapter Nine.

OTHER EQUIPMENT

Other than the items previously mentioned, the equipment you will need for your new cat is minimal.

Generally speaking, a bed for a cat is useless because a cat will sleep wherever she wants. When you first bring a kitten or cat home, it is nice to provide her with an old towel, sweater, or afghan. Place it in a nice warm spot, away from household confusion, and show it to her. Chances are she will curl up on it for a while. But as soon as she becomes used to your home, she will sleep wherever she chooses. And she may very well choose a different spot at different times of the day, on different days, in different seasons, and seemingly just at whim. During the night, a cat that is truly bonded to her owner may sleep on the owner's bed.

One or more water bowls should be available at all times. They needn't be special water bowls, but it is a good idea if they are heavy enough not to tip easily and are easy to clean. (See "Cats and Drinking Water," pages 105–6.)

By the same token, no special feeding bowls are needed. Any plate or bowl will do—some owners feed their cats from paper plates or towels. If you plan to leave dry food out all the time, a weighted bowl, or one with a nonskid bottom, is best so it cannot be pushed off a countertop. (Note: If you have a dog, you will have to feed your cat on a countertop or table, or the dog will quickly eat her food.) I do not recommend combination food/water bowls—the food gets into the water and vice versa.

You should also have a comb and brush on hand. Despite their constant self-grooming, cats need daily brushing and combing to remove loose hair and help prevent the formation of hairballs from ingested fur.

Special toys are not necessary. A crumpled piece of paper, a paper towel or toilet-paper roll, an empty paper bag or box, and a dangling string can all provide endless entertainment for a kitten or cat. But most cats do like catnip, and a catnip mouse or other catnip toy can be a lot of fun for a cat and her owner. So can mechanized mice (although some kittens are afraid of them) and feathery objects that dangle from a stick.

A PLACE OF HER OWN

If your new kitten or cat is an "only child" coming into a home without any other pets or children, she should certainly have the run of the house. But if she is coming into a household with other pets and/or small children, she may need a gradual introduction. This is especially true if she has never lived in a household filled with other pets and children. Obviously, a kitten or cat that comes from a lively household is better equipped to deal with the same situation. Your common sense will tell you whether or not your new pet will require a gradual introduction into your household or is able to plunge right in.

So far, I have talked mainly about the day-to-day equipment you will want to have on hand when you bring your new pet home. But you may also want to provide your kitten or cat with a safe haven (especially if you have pets or small children), at least until she becomes used to your household and its inhabitants.

A crate can be a good solution. It should be large enough to hold a small litter pan, water and food dishes, and a comfortable place to sleep such as a carpeted shelf. It can be placed in a corner, out of the way of household traffic. It should be in a location where the new kitten or cat can see all of the activity going on but not feel threatened by it. Other people or animals can approach the crate without scaring her. She needn't stay in the crate all of the time, but it can be a sanctuary for her as she becomes used to the household. Eventually you can leave the door of the crate open so that she can move around but will still be able to retreat to the safety of "her" place when she needs to.

Alternately, if she seems afraid only of a dog or small children, a pressure gate across the door of a small room can provide her with a sense of security. She can jump in and out at will, while the dog or children cannot. Again, provide her with a litter pan, food, water, and a comfortable place to sleep. As she becomes used to the other members of the household, she will spend less time in her room, and usually the gate can (eventually) be removed.

Chapter Six

A Happy, Healthy Cat

Now that your new pet has settled in to your home, you should pay attention to her care and health needs. One of the most common and unfortunate myths about cats is that they are virtually "no care" pets. People often think that cats require little in the way of attention or health care; that they are not as needy or dependent on their owners as dogs and can be left alone for long periods of time with no ill effects.

If you really want to have a good relationship with your cat and to bond with her, and possibly go on to train and interact with her in a meaningful way—in other words, to indulge in CatSpeak—you will pay particular attention to her well-being. It will take *commitment* on your part to bring your relationship one step further.

Your pet cat is willing and able to give you unquestioning affection, warmth, and companionship, to say nothing of amusement, playtime, and constant amazement. All she requires of you is safety, a warm place to sleep, a diet that fulfills her body's needs, fresh drinking water, appropriate medical care, some entertainment or distraction, *and* consistent loving attention and interaction with you.

I like to think of appropriate cat care as encompassing the Three C's: Compassion, Care, and Concern (just as I have dubbed my training philosophy the Three P's: Patience, Persistence, and Praise). One of the things I think a cat owner must do is to put himself in his pet's paws, and try to realize what the cat feels and observes in order to truly engage in CatSpeak and understand where his cat is "coming from."

With that in mind, let me discuss some health and care musts to follow if you are going to have a happy, healthy pet cat.

YOUR CAT'S DOCTOR

I discussed how to choose a veterinarian and what to expect on the first visit to the doctor in Chapter Five. Now I want to discuss ongoing care.

Most veterinarians maintain records of their patients (or clients) and will send reminders when booster shots or annual physicals are due. It is important to respond to these reminders. Immunizations against infectious and contagious feline diseases require regular boosting to remain effective, and there are a number of important protective vaccinations that your cat needs to stay healthy. Even if she is an indoor-only cat, she still must be protected against such virulent diseases as rabies (even a bat that comes down the chimney or through an open window can infect her; also, a rabies certificate is required in many states). She also needs protection from other feline diseases that may be airborne or carried on people's shoes and clothing. Your veterinarian will be able to advise you what immunizations are called for in your particular geographical area.

An annual physical is important. When a veterinarian sees your cat on a regular basis, he will be able to spot abnormalities or changes that you may not have noticed. An office visit also provides you with the opportunity to discuss any change in your cat's physical condition or behavior over the past year.

Of course, special circumstances such as an illness, infection, or urinary tract disorder may require more frequent visits. Once your cat becomes a senior citizen (from twelve to fifteen years old) you may want to have more regular exams, especially if she suffers from some ongoing condition such as diabetes, liver, thyroid, or kidney problems, arthritis, or a gradual loss of hearing or eyesight.

Bear in mind that a cat, unlike a dog, will not run to you for help if she is ill or has been hurt. If a cat is not up to par, her natural instinct is to protect herself from predators. She will find a small, dark, isolated spot and hide.

She also won't tell you if she isn't seriously ill but simply is not feeling well. The only way you will know all is not right with your cat is if you notice a change in her habits. Any unusual behavior that lasts more than a day or two may be a signal she is not feeling well. If she suddenly stops asking for breakfast first thing in the morning, or doesn't seek out her usual sleeping spot on your bed at night, you should suspect something's wrong. This is the time to seek veterinary help, and ideally, you've established a good relationship with a veterinarian you can trust and talk to easily. This step is most important in maintaining your cat's good health.

Neutering

Throughout this book, I will stress again and again how important it is for your pet cat to be neutered. Neutering consists of an ovariohysterectomy, or spaying, for females, and castration, or altering, for males. Not only does neutering at the appropriate time prevent fatal diseases such as cancer, it will also lessen or eliminate many behavior problems, as well as accidental breeding that adds to the tragedy of unwanted kittens.

Who needs a female cat that regularly comes into heat every ten days or so and paces restlessly, howling and "calling," throughout the day and night? Or a male cat that moans and howls, tries to get out, becomes aggressive, and sprays urine all over the house? Neither of these cats is an ideal pet.

It used to be recommended that a cat not be neutered until he or she reached sexual maturity, at around six to eight months of age. But some shelters have recently begun to neuter their kittens *before* they are adopted, at a much earlier age. This is designed to prevent the birth of more unwanted kittens. The long-term health effects of early neutering have not yet been determined. Therefore, if your cat has not been neutered when you adopt her, it is probably best to wait until the recommended age. Consult your veterinarian about this.

ALTERNATIVE THERAPIES

Veterinary medicine has embraced alternative forms of therapy (just as human medicine has) when traditional medicine doesn't seem to do the trick. For example, cats can suffer from many forms of painful arthritis and may be helped by acupuncture or massage.

Acupuncture

The American Veterinary Medical Association (AVMA) recommends acupuncture for many painful cat problems. Veterinarians may prescribe acupuncture, along with appropriate medications, for feline patients that suffer from painful arthritis, degenerative joint disease, paralysis, epilepsy, and asthma.

Acupuncture began in China thousands of years ago and is still the only treatment used by much of the world's population. It is based on the principle of *chi*, an energy system that flows along meridians through the bodies of all living creatures. When this energy system is blocked for any reason, it can cause pain, disease, or illness.

An acupuncturist aims to open the blockage by inserting needles precisely along the exact meridian that leads to the painful or diseased area(s) in the body. This is meant to release certain neurotransmitters and neurohormones, such as endorphins. Endorphins are amino acids that act as natural painkillers and can help a cat in pain feel better.

If you opt to use acupuncture to help your cat overcome pain, make certain you will be allowed to be present to soothe and reassure her (CatSpeak to her) so she will not be frightened.

Massage

Massage is somewhat akin to acupuncture, because it also releases endorphins when practiced correctly. Everybody loves a massage, so why should your cat be any different? Performed properly, massage therapy will relax her muscles, calm her, and make some of her aches and pains go away. Even if she is not in pain, a good massage is a wonderful way to strengthen the bond between you and your cat and enhance her trust in you. It will calm an overactive cat, make an ailing cat feel better, and in general, make you both feel more mellow.

I have developed my own (nonscientific) massage for cats.

Bash's Magical Massage for Cats

- Start at your cat's head and firmly and gently massage her skull with your thumbs. Begin at her nose and work your thumbs gently in a circular motion along her cheeks and then behind her ears and down the sides of her throat. Move upward onto her ears and gently stroke and "pull" her ears between your fingers. If your cat is uncomfortable when you touch her ears or the area around them, you should suspect an ear infection or inflammation and seek veterinary help.
- Next, move to your cat's neck and run your thumbs firmly across her shoulder blades and slide them down her spine. Use your thumbs to move in a circular motion and exert some pressure on her spine.

- Move on to her legs (which she may not like at first, but have Patience, go slow, and she will come around). Rub and stroke each leg separately with your thumbs, starting at the cat's body and working toward her paw. End up massaging all four paws, paying particular attention to each toe pad.
- Once you and your cat have experienced my cat massage, neither of you will ever again be content with simple petting.

Homeopathy

The AVMA reports that homeopathy has had wonderful results, particularly in cases where conventional medicine has not helped.

Homeopathy was first practiced by Samuel Hahnemann in the early nineteenth century in Germany. It is based on the belief that diseases or illnesses can be successfully treated by administering minute doses of a substance that creates symptoms similar to those produced by the disease. Homeopathic remedies are usually made from plants and plant products, but some are made into capsules or tablets from highly diluted animal or mineral extracts.

Homeopathy is still not recognized by much of the scientific community, but if your cat is not getting better with traditional medicine, it is worth asking your veterinarian about it.

Other Therapies

Other alternative therapies that are sometimes used to treat cats are *chiropractic therapy* and *neutraceuticals.*

Chiropractic medicine is based on the belief that abnormal tensions on the spinal vertebrae cause nerve irritation. Proper manipulation of the spine is intended to relieve this irritation. This therapy has not been approved by the AVMA, but some veterinarians recommend it in certain cases.

Nutraceuticals are dietary supplements formulated to prevent or treat disease. They are sold in pet-supply stores and over the Internet, but I do not recommend giving them to your cat without a veterinarian's approval.

WHAT'S FOR DINNER?

Cats have a reputation for disdaining food, or at least for not being as food-oriented as dogs are. This may be true of some cats, but most I've known (and I've known a lot) really like to eat. True, some cats prefer some kinds of food, or one particular food to another. But most cats enjoy their food.

There are three main kinds of cat food on the market: moist canned, semimoist, and dry. When you adopt your cat, find out what brand and flavor of food she is used to eating, if possible. A sudden diet change can upset even the hardiest cat's digestive tract. If you want to change her diet later on, do it gradually. Mix a little of the new food into the old, increasing the amount a little each day. If your cat's stomach becomes upset, reduce the amount of new food and begin again.

Cats have several specific dietary needs. They require vitamin A in their diets but cannot manufacture it in their own bodies. They also need a daily source of niacin and taurine. That is why it is important to feed a cat food designed for her (a kitten needs kitten food to fill her growing needs, for instance). It doesn't really matter what form the food is in as long as it is "balanced and complete, for cats." Many owners feed their pets a variety of foods. Dry food can be left out for snacking, while canned is saved for dinner, breakfast, or both, and semimoist for treats. (Note: Semimoist foods contain preservatives, and some veterinarians do not recommend them as a sole diet for cats.)

Most cats have healthy appetites. Some are always ravenous. If yours is one of the latter, be careful not to overdo. A fat cat is not a healthy cat. She is putting strain on her heart and other internal organs. If you look at your cat from above, you should be able to see a slightly defined waistline just in front of her back legs. If you cannot, she may be too fat. Your veterinarian will know right away if your cat needs to lose weight and can advise you about her diet.

There is also the well-publicized finicky cat who

turns her nose up at everything you offer her. As I mentioned in Chapter Two, this supposedly finicky behavior might be due to the simple fact that a cat's nose is chronically stopped up and she cannot smell her food. If you think this is the case, take her to the doctor, who can easily solve this problem.

If a cat is under stress she may not want to eat until the cause of the stress goes away or she becomes accustomed to whatever it is (a new pet or person in the household, for instance). It won't do most well-fed house cats any harm to miss a meal or two, or three, or more.

Most owners like to feed their cats a variety of foods and flavors. Fish one day, chicken the next, followed by beef, and so forth. But some cats really like only one kind, or flavor, of food. As long as the food is complete and balanced, there is absolutely no harm in feeding your pet the same food day after day. You may think it's boring, but if your cat continues to eat it, she obviously doesn't.

People food is not good for cats. For one thing, it does not provide the nutrients a cat needs to be healthy; for another, it is very likely to cause digestive problems. This doesn't mean you shouldn't give your cat a scrap of chicken or fish from your plate as a treat. Just don't fall into the trap of giving your cat nothing but human food. If you encourage or allow her to develop a taste for tuna meant for people, for instance, your cat will not thrive, because she is not getting the nutrients she needs. What's more, an all-tuna diet that is not properly supplemented for cats can cause inflammation of a cat's body fat. Don't ever give her people tuna and she will never know the difference!

There are special prescription diets designed to meet specific cat health needs (heart or kidney disease, for instance). They are available only through veterinarians and usually come in only dry and canned formulas. Most cats learn to like them.

Mealtime should be a happy time for you and your cat. When you feed her, stay nearby while she enjoys her food. Talk to her, stroke her, and tell her she is a good girl. This will bond you with your cat in a pleasant way and establish another CatSpeak opportunity.

A Preference for High Places

DOMESTIC CATS USUALLY PREFER TO BE OFF THE GROUND WHEN they eat and drink. Perhaps the eating behavior is a holdover from leopards, which always take their killed prey up into a tree to prevent marauders such as hyenas from stealing their food.

Cats like to eat on a chair, countertop, or table, where they can easily see around them. I always feed my cats on a counter because I also have dogs that would immediately scarf up the cats' food if I put it on the floor. At one point, all of my dogs were away and I put the cat food on the floor. Although the cats approached it with interest and sniffed at it, they did not actually bend down and eat. I figured out that they felt too vulnerable if they put their heads down to concentrate on eating while they were on the ground. They wanted to see around them while they ate. I moved their food back up to the counter, where they immediately ate it.

By the same token, most cats do not like to drink water from bowls on the floor. As I mention in "Cats and Drinking Water," below, most house cats I have known prefer to drink from a dripping faucet or water fountain than from a bowl on the floor.

Cats and Drinking Water

A CAT REQUIRES DRINKING WATER, AS ALL MAMMALS DO, OR she will become dehydrated and may develop kidney or bladder problems that can make her very sick. This is especially true when a cat eats only dry food that contains no moisture. If she eats canned food, she will get some liquid from it, but not enough.

Cats are often very fussy about the water they drink. Most cats will not drink from a bowl of water that has been standing long enough to form a light scum on the top; nor will they appreciate sharing a drinking bowl with a slobbery dog. They really like fresh water!

An owner can solve this dilemma several ways. When a cat wants a drink (she will usually indicate this by standing by the water bowl, looking at you, and meowing), rinse the water bowl, dry it out to remove any film or scum, and fill it with fresh water.

Many cats like to drink directly out of a faucet if it is left on in a gentle stream or drip and will jump up on a countertop or basin and "ask" for water. Some cats will lean into the toilet and drink fresh water from the bowl if the lid is left open. (If your cat does this, don't use cleaning chemicals.) Other cats really love to go into a shower or bathtub as the water drains out and lap water off the floor. The same caution about chemical cleansers applies here.

Some owners provide their cats with constantly running, decorative drinking fountains.

Despite their usual fastidiousness about drinking water, many cats develop a habit that should be discouraged—they drink out of plant saucers. This water may contain potentially dangerous minerals and chemicals from the plant soil and the pot itself that can upset the cat's digestion.

Milk can be a suitable substitute for water, but many cats cannot digest whole cow's milk and get diarrhea when they drink it. Sometimes diluting milk with water may help. There are milklike products on the market, specifically formulated to be more digestible for cats. Some cats prefer them to water. Pedialyte, designed for infants, contains electrolytes and is an excellent source of needed liquid for a cat, especially when she is not feeling well.

WHAT A BEAUTIFUL PUSSY YOU ARE!

It is extremely important to groom your cat on a regular basis. Not only for looks, which are certainly important, but also to remove shed hair and for healthy skin and good circulation.

All cats shed continuously in a natural cycle of hair replacement that has to do with both the photoperiod and environmental temperature. A cat that is under stress will always shed a lot. If shed hairs are not removed on a regular basis, a cat will swallow them when she grooms herself and may form hairballs in her stomach, which she will then spit up while making horrible hacking and yowling sounds. Even well-groomed cats may develop hairballs. Veterinarians sell oral preparations that help prevent the formation of hairballs. If your cat suffers from hairballs, ask your doctor about these remedies. When longhaired cats are not groomed regularly, their shed hairs will form mats and tangles.

A grooming session provides an excellent opportunity to look your cat over. It is also a wonderful bonding process. It doesn't make any difference where you groom your cat. If she does not like to be held, a table or countertop is a very good grooming surface. It is a good idea to assemble whatever grooming tools you will need and keep them handy in a drawer, box, or basket.

Grooming Tools

- Brush or comb, or both
- Nail clippers
- Cotton swabs
- Blunt-nosed scissors for longhaired cats
- Cat toothbrush and appropriate toothpaste
- Framed screen for bathing (more about this under "A Bath for a Cat?", pages 111–13.

Brushing and Combing

Both longhaired and shorthaired cats can be brushed or combed. Combing a longhaired cat every day will prevent mats from forming in her soft undercoat. If mats do form, the best way to remove them is to pull them apart gently with your fingers, and then out. Don't try to cut them out because it is very easy to pinch or cut a cat's skin. Go to a professional groomer, or have the veterinarian deal with a badly matted cat.

If the fur flies when you comb or brush your cat, a slightly damp cloth or special wipe rubbed over the cat's coat after grooming will help capture the loose hairs.

Eyes and Ears

Cats' eyes usually remain clean, but if matter should form in the corners alongside the nose, wipe it gently away with a damp cotton swab. Longhaired cats may need to have excess fur around their eyes trimmed a bit with blunt-nosed scissors. Be very careful when you do this! If your cat is at all skittish, have someone else hold her while you trim around her eyes, or go to a professional groomer.

Look in your cat's ears while you are grooming her, and if you notice any wax or dirt, remove it with a moistened cotton swab.

Mouth and Teeth

It is a good idea to examine your cat's mouth and teeth on a regular basis. Pull back her lips and look at her gums. They should be pink and firm. If they are pale or bleed when you touch them, or if there are any swellings or sore-looking spots, take her to the veterinarian.

Cats' teeth and gums need regular care, just as people's do. They should be brushed on a weekly or biweekly basis. If you start your kitten off early so that she becomes used to having her teeth cleaned, you should have no problem. Older cats may take a little time getting used to it, but the process is so beneficial that it really is worth it.

You can obtain a kit with everything you need to clean your cat's teeth from your veterinarian or pet-supply store. Or you can simply wrap some gauze or a washcloth around your index finger to use as a "brush." Some owners make toothpaste of baking soda, salt, and hydrogen peroxide, but your cat will probably prefer flavored kitty toothpaste. Also, toothpaste that is made for cats does a better job of breaking down plaque.

Nail Clipping

Nail clipping is a very important part of grooming and should be done every few weeks. Again, if you start your cat off early, she will probably not object to nail clipping. If not, you may run into some resistance.

Claws that are too long will catch on things. If a cat becomes stuck on a piece of furniture, she will not only damage the furniture, she may hurt herself trying to get loose. A cat with overlong claws may also accidentally scratch another pet or a person. In addition to trimming, a good sisal or rope scratching post will help maintain your cat's claws.

If you have never clipped a cat's claws, ask the veterinarian to show you how. You have to be very careful not to clip too far, or the nerve and blood vessel (the "quick") in the nail will be cut. This will hurt the cat and will bleed a lot. It will be very difficult to get your cat to stay still for nail clipping after that!

To get a cat's claw to extend, press gently on the bottom of a footpad with your thumb while you hold her foot. If your cat does not like to be held, try sitting her on your lap, rear against your body.

If your cat is especially uncooperative, you may need to wrap her in a towel and release one paw at a time until she learns to relax and trust you. To cut a young, wiggly kitten's nails, have a helper hold her up in the air by the nape of her neck, legs dangling, just the way a mother cat would restrain a kitten.

A Bath for a Cat?

Here's another cat myth I'd like to debunk: Cats don't like baths. I bathe my cats often, and they really enjoy the experience. A warm, calm, soothing bath can be a therapeutic experience for a cat as long as her owner can learn to relax and enjoy it. Remember another of my favorite sayings: "A nervous owner makes a cat nervous."

Longhaired cats benefit especially from regular baths that help prevent their fur from matting. Show cats are bathed all of the time to keep their coats clean and shiny. If you or anyone in your household is allergic to cats, regular baths will significantly cut back on the amount of saliva on their fur. (Note: We now know that it is the cat's dried saliva on her fur, not the fur itself or its dander, that causes people to suffer from allergic reactions.) There are also some cats that exude matter. Wherever they sit or sleep, greasy black stuff seems to accumulate—a bath would be very good for them. Although indoor cats rarely get fleas unless another animal (such as a dog) introduces them into a household, a cat with fleas will have to be bathed. Use special hypoallergenic shampoo, which will not dry out the skin.

If your cat has not been bathed before, or if you are at all uncertain, get someone else to help you. Your cat's first bath experience should be relaxed and calm, and this will not be true if you are frantic, nervous, or unsure of yourself. A calm person who is familiar to the cat can hold her, help reassure her, and make the entire process more pleasant.

Assemble all of the equipment you will need. I recommend that you place a framed screen in the bottom of the tub on top of a rubber mat, so the cat can hold on and not slip. You can assemble the screen yourself: Measure your tub and build a frame to fit out of strips of wood, then staple a piece of screening onto the frame. Be sure the staples are on the bottom so the cat cannot loosen them.

Use shampoo designed for cats and follow the instructions—some shampoos need to be diluted. Have a washcloth and several large towels nearby. It is also a good idea to protect a cat's eyes and ears from soap and water. Place a drop of mineral oil in each eye, and put some lamb's wool or cotton in each ear.

Try not to wet the cat's face too much. Use a damp cloth to wash around her eyes and muzzle. Avoid spraying her face with water and don't, by any means, pour water over her head or submerge her head underwater.

Sometimes cats are spooked by running water, so I recommend filling a tub with warm water beforehand. Place the cat in the tub, wet her with a washcloth, or pour water over her body, starting at her neck. Rub shampoo into her fur (don't use too much shampoo or it will be hard to rinse out). Massage the shampoo into a lather all over her body as you talk to the cat and reassure her—you can make this into a mini-massage. Then rinse thoroughly with warm water to be sure you remove all soap residue, which can cause skin irritation.

Wrap your cat in a large towel and rub her dry. With my longhaired cats, I use a blow-dryer with the heat and speed at the lowest setting while I comb them dry to help keep tangles from forming. This may take your cat some getting used to. If you are going to use a dryer, put the cat on a counter or tabletop to blow her dry; otherwise, rub and comb her, then be sure she is in a warm, draft-free location until she is thoroughly dry.

Congratulations! You have just successfully given your cat a bath.

EXERCISE AND ENTERTAINMENT

I assume you have chosen to keep your cat indoors, for reasons of health and safety. It is a scientific fact that indoor-only cats lead significantly longer lives than indoor-outdoor or outdoor-only cats.

But indoor-only cats may suffer from boredom and inactivity. They are likely to become sedentary and overweight. How to offset these problems? Some owners opt to create a compromise. They provide a screened-in area so the cats can be outdoors and still be safe. A window perch or climbing post next to a window can also provide a wonderful vantage point for a cat to look out at squirrels, birds, butterflies, rabbits, other cats, people, and dogs walking by. Some cats also enjoy "kitty videos," tapes of birds, fish, gerbils, and so forth.

Although all of these things provide entertainment, they do not provide an indoor cat with the chance for exercise.

When cats are young, they usually need no encouragement to jump, climb, run around, and explore. As they become older, cats regularly sleep for ten or more hours a day. If their sleep is not interspersed with some activity, lethargy (and more sleep) will soon set in. One way to help offset this is to provide distractions such as scratching posts, climbing furniture, and toys. But after a while, the novelty wears off and most cats will eventually ignore these enticements. Other pets in a household provide some entertainment and distraction, but only sporadically.

Even when there are plenty of toys, furniture, and other pets in your household, your cat still needs *your* daily interaction with her. It takes only a few minutes for you to toss a Ping-Pong ball or piece of crumpled paper for your cat to chase and fetch. Dangle a string, or put an empty paper bag or carton on the floor and see your cat go crazy for a while! Give her some catnip on the counter or table, and watch her roll in it. She will probably then go into a "silly fit" and run all over the house. This is great exercise.

I am a firm believer in interaction with your cat on more advanced levels. In Chapter Eight, I describe several basic training steps you and your cat may enjoy; and in Chapter Ten, I talk about even more things you can do with your cat to help prevent her from ever becoming a bored couch potato.

THE SENIOR YEARS

Genetics, care and environment, stress, and disease all combine to determine the time when your cat becomes a "senior." Signs that a cat is getting old can be both behavioral and physical. When your cat becomes older, she will probably slow down physically, just as we all do. She may sleep more, begin to snore and sneeze a little, become less agile, and be more set in her ways.

She may lose some teeth; her eyes may become cloudy and less colorful, and her coat may loose its shine. An old cat may lose weight, and her hips and backbone may appear more prominent.

She may also suffer from some stiffness and joint pain. As I said before, a cat will not "tell" you she is in pain (unlike a dog). You will have to discover it for yourself. If she begins to hesitate before she leaps up on the countertop to get her dinner, or if she chooses to sleep on the floor by your bed instead of up on top of it, you may assume that it hurts her to jump. You can provide her with some help. A chair next to the kitchen counter or a hassock at the foot of the bed will provide her with steps so she can get up where she wants to without pain or embarrassment.

An observant owner will notice any change in his cat. Regular grooming is especially important with an older cat. Any lumps or bumps require immediate veterinary assessment. Early detection and treatment of any abnormality is particularly important for older cats because their resistance is not as strong as it used to be, and the risk of serious illness can be greatly diminished by quick action.

This is a time when it is especially important to interact with your cat on a daily basis. Reassure her all the time with CatSpeak. Let her know you really love her, care for her, and want her to feel good. Encourage her to interact with you through petting, stroking, massaging, and all the other good ways of communicating that you have always enjoyed with each other. Do not ask her to do anything that might be painful or dangerous for her at her age—in other words, don't ask her to jump through hoops (see Chapter Ten) if she seems to be in pain or unsure of herself.

An older cat who has lost teeth may not be able to eat dry food anymore. Give her a soft canned diet. If she has always liked dry food, soften some in a bit of milk or water so she can still enjoy it.

As she becomes thinner and bonier, an old cat often feels cold, and she will appreciate softness and warmth. For my clients with senior cats, I often recommend a heating pad covered with a blanket or soft towel as a wonderful place for a senior cat to relax and be comfortable.

A TIME TO SAY GOOD-BYE

Perhaps your old cat will die peacefully in her sleep. This would be nice. But more often, you will be faced with a decision. At what point does the quality of life of your beloved pet become so bad that you are forced to make the choice to end her life?

Hopefully, you will have established a good relationship with a veterinarian who can help you in this difficult time. Is there any hope that your cat will feel better and be able to enjoy her life and her relationship with you again? Are there heroic measures that can help her, or is she going to continue to fail, feel terrible, and be unable to participate in life as she used to?

These are all questions you must come to terms with. If you conclude that it is not doing your cat a kindness to keep her alive, you will probably come to the unhappy conclusion that euthanasia is the best answer. You don't want your beloved cat to suffer anymore. It is time to stop her pain.

Almost all of us will outlive most of our pets. But how do you deal with the loss of a beloved pet? If you have other pets, they may provide some solace (although they will also miss their friend and not understand her sudden absence).

People may say, "Oh, it was only a cat, get over it," or "Get another cat." These are people who do not understand the depth of feeling you can have for a particular cat and the unconditional love you have lost. They don't know about the wonderful communication (CatSpeak) you have had with your pet.

In time you will heal. If you have real problems coping with your loss there are professionals who can counsel you: Seek their help. Time will heal, and I am willing to bet that eventually you will have another delightful feline to love, bond with, train, and trade CatSpeak with.

This little "fairy tale" is a favorite of mine. It may help you and your family cope with your loss.

The Rainbow Bridge

Just this side of heaven is a place called the Rainbow Bridge. When an animal that has been especially close to someone here dies, that pet goes over the Rainbow Bridge. There are meadows and hills for all our special friends so they can run and play together. There is plenty of food, water, and sunshine, and our friends are warm and comfortable.

All the animals who had been ill or old are restored to health and vigor; those who were hurt or maimed are made whole and strong again, just as we remember them in our dreams of days gone by. The animals are happy and content, except for one small thing: They each miss someone very special to them who had to be left behind. They all run and play together, but the day comes when one suddenly stops and looks into the distance. His bright eyes are in-

tent; his eager body quivers. Suddenly he begins to run from the group, flying over the green grass, his legs carrying him faster and faster.

You have been spotted, and when you and your special friend finally meet, you cling to each other in joyous reunion, never to be parted again. The happy kisses rain upon your face; your hands again caress the beloved head; and you look once more into the trusting eyes of your pet, so long gone from your life but never absent from your heart.

Then you cross the Rainbow Bridge together. . . .

—Author unknown

Chapter Seven

The Vocabulary of CatSpeak

Cats communicate easily with one another by scent-marking (which I discussed in Chapter Two), vocalization, and body language. If a cat accidentally wanders into another's territory, she immediately knows from the other cat's body language and vocalization whether or not she is going to be welcomed, tolerated, or chased away angrily. This understanding is important, as it can often help avoid a serious fight. Although unaltered males, or tomcats, do often engage in bloody battles over females, it rarely leads to death because the loser indicates with body language that he is willing to back off.

Cats always understand such signs and act appropriately; just a lowering of the head or the swish of a tail are clear feline signals that every cat, from a lion to a tabby, can easily comprehend.

A human may not understand his pet cat's signals. This lack of communication can create misunderstandings that cause the owner to react in ways the *cat* does not understand. She may then be judged "bad," "aloof," or "antisocial," which can result in her ultimate abandonment. This is a really sad situation, and one that can almost always be prevented if an owner is willing to learn more about his pet: what her special qualities are and how she communicates her feelings and needs.

In this chapter I hope to show you how to interpret your cat's behavior by learning her language. In order to learn any language, you must first learn the vocabulary. Here I will talk about the vocabulary of CatSpeak.

I talked about cats' sensory abilities in Chapter Two and explained how they far outreach those of humans. Therefore, cats' reactions to their environment, and to the people, things, and other pets around them may be a lot more exaggerated than human reactions.

A cat communicates her feelings with a combination of signals, using her face, ears, body, tail, paws, and voice. Therefore, you need to look at the total picture in order to interpret how your cat is feeling and what she is trying to tell you.

FACIAL EXPRESSIONS

A cat's face, just like a human's, is the most expressive part of her body. But unlike humans, who use only their eyes and mouths to express their feelings, cats use every part of their heads. They use their ears, whiskers, mouths, and eyes to try to communicate their feelings about you and what you are doing. For example, it is very important to know that narrowed eyes and flattened ears and whiskers, accompanied by a swishing tail and a low, growling noise, means "Stop what you are doing right now, or I am going to have to bite or scratch you to make you stop!" Your cat feels she has given you fair warning, and if she does bite or scratch, she is not being bad or mean! She is simply protecting herself from further hurt or discomfort.

What Whiskers Say

Most animals have whiskers, but cats' whiskers are especially versatile. In addition to enabling cats to "measure" spaces so that they do not become stuck, cats' whiskers are very expressive. They communicate everything from happiness to anger and all the other emotions in between.

In a normal position, a cat's whiskers are held together very loosely, so that they fan straight out at the sides of her face or slightly toward the front. This indicates that she is relaxed, content, and receptive to attention.

An aggressive, angry cat.

If a cat's whiskers are pointed backward and flattened against her face, it means she is either annoyed and angry or afraid of something and adopting a defensive posture. You may see your cat's whiskers flatten if she is suddenly confronted with a strange dog, cat, or even human; or if she hears a harsh or unusually loud sound.

Some cats are more fearful than others and may run and hide, whiskers flattened, at the slightest sudden noise or commotion. A thunderstorm, children running and playing in the next room, a doorbell ringing, loud laughter or conversation—all may trigger a fearful response in a "scaredy-cat." If you understand by observing her whiskers and body posture that she is really afraid, the

A relaxed and alert facial expression.

best thing to do is to leave her alone and let her hide. In time, she may become used to activities that once frightened her, but it is useless to try to reason with a frightened cat. Attempts to drag her out of her hiding place will only scare her more and will probably result in your being scratched or bitten.

A cat that is excited will often fan her whiskers out toward the front of her face.

Cats use their whiskers in a unique way to explore their environment—like antennae. When a cat's whiskers are fanned out to the sides of her face and slowly moving around in circles, it means she is very curious about something in the area around her and is using her whiskers to "feel" the air currents and figure out what is happening or where people or other animals are. She may do this often; for instance, if someone enters a room or if a window or door opens. A cat's vibrissae play a very important part in her assessment of the world around her.

What Ears Say

If only we could move our ears the way cats can to indicate how we are feeling! Along with whiskers, the position of a cat's ears speaks volumes to an observant and knowledgeable owner. Most cats' ears are pointed or slightly rounded at their tips and sit straight up on top of their heads. Notable exceptions are Scottish folds and American curls, whose ears are bent. But even with these variations, a cat's ears remain very expressive.

In a normal, relaxed position, a cat's ears are turned forward and perked straight up. A cat with perked ears is content. If she is curious about something, she may swivel her ears sideways, turned toward the outside, so she can better capture whatever odd or interesting sound she hears. (You may not hear anything at all. Remember that cats' hearing is many times more acute than humans'.)

When a cat is afraid, her ears will be pulled backward and flattened against her head, just as her whiskers are. Perhaps she is trying to make herself appear smaller to the "enemy." Or maybe she is performing the equivalent of what a child does when he puts his fingers in his ears to block out a frightening noise. Whatever the reason, leave a frightened cat alone. If you try to comfort her, you will probably be treated aggressively. Her fear and sense of self-preservation will almost certainly overcome any normal sense of security she feels with you.

If a cat's ears are flattened backward, it is a sign of anger. Accompanied with flattened, pointing whiskers, this is a clear sign that the cat is seriously upset. Keep your hands and face at a safe distance!

If a cat's ears are held flat out to the side, it most often means that she has an irritation or pain in her ears. This behavior is usually accompanied by ear scratching. If your cat consistently carries her ears in this position and scratches at them often, seek veterinary help. This ear position may also indicate that she is not feeling well in some other way.

What Eyes Say

Cats use their eyes to communicate their feelings, as all animals do. Some cats' eyes are round, some slightly slanted. Others have protuberant eyes. But in general, all cats' eyes express the same feelings. All cats (with the possible exception of Siamese, which are often cross-eyed) have binocular vision; that is, they are able to see objects in depth and perspective, just as humans can. Their vision is keen, especially in low light. Cats' pupils are vertical, expanding and contracting according to the amount of light: In low light, large, dilated round pupils fill up most of the eye; in bright light, the pupils are very narrow, slitted, and vertical.

When a cat is in a relaxed state of mind in average room light, her eyes will be open but not stretched widely, and her pupils will look normal, neither exaggeratedly dilated nor slitted. If she is in bright light, her eyes will normally be open with the pupils narrowed. When she is content but somewhat drowsy, her eyes may close partially or all the way.

A calm and happy cat will often blink her eyes slowly while looking at her human in a kindly way. Cats often stare unblinkingly at their owners, which can be disconcerting. Usually, a cat stares at you to discover what you are doing or are going to do next. Cats do not seem to notice or care if you stare back at them, as opposed to some aggressive dogs, which may read staring into their eyes as a challenge.

Although cats see amazingly well in low light, they often take awhile to focus. When you approach a cat from across a room, she may stop and stare at you in a rather eerie way, as if she has seen a ghost. This is simply because she has to take a moment for her eyes to focus until she recognizes you. She is concentrating on your appearance and waiting for you to move as she attempts to sort out your image. Siamese cats, in particular, often indulge in a gesture that seems as if they are looking through bifocal eyeglasses. They raise their heads up, look out of the bottoms of their eyes, and seem to be adjusting their line of vision.

If a cat is feeling just a bit irked, she may narrow her eyes a bit. But narrowed eyes and pupils, along with flattened ears and whiskers, indicate extreme anger. Watch out!

An annoyed cat.

Very wide-open eyes with dilated pupils can mean a couple of different things in cats. They may simply indicate that the cat is attempting to see as much as possible as quickly as possible in a new or unfamiliar location. For instance, if she is in unfamiliar surroundings, such as the veterinarian's office, or in your home for the first time, she may be trying to take it all in as quickly as possible. A cat's pupils also dilate when she feels threatened and is close to doing something about it. In other words, this signal can be a warning that she is about to strike out either in self-defense or anger.

Eye Contact

One of the many ways to bond with a pet cat is to make visual contact with her. When your cat stares directly into your eyes, look back with interest and fondness. It may be difficult to tell for sure, but it seems evident to me that this kind of eye contact does a great deal to assure your cat that you care for her and understand, at least a little bit, what she is trying to communicate to you.

What Mouth Position Says

In addition to vocalization, which I will talk about later in this chapter, a cat's mouth itself is expressive. The famous Cheshire cat in *Alice in Wonderland* is not the only cat that smiles. Domestic cats, when content, looking at their owners, and purring, often seem to smile—at least the corners of their mouths do turn up, giving them a happy face.

Pulled-back lips and a slightly open mouth probably mean that a cat is engaged in flehmen, a response to certain odors. I talked about cats' highly developed sense of smell and this particular reaction in Chapter Two.

Quite naturally, you will be able to see that your cat is very angry if her mouth is open and her teeth bared.

BODY LANGUAGE

Along with her facial expressions, a cat's body is a clear indicator of her mood and feelings. Just the way she stands or sits should let you know if she is relaxed and calm, frightened, or angry.

When a cat greets you (or even a friendly dog or other cat) happily, she will curve up her back and rub her body against you. Her tail will be straight up in the air and she will usually also butt her head against you and wind her tail around your legs. This is a sign of affection. It also means that the cat is marking you as her property with the scent glands located in her cheeks and on top of her tail.

Sometimes a cat will flop belly-up on the floor in front of you, seeming to want her tummy rubbed. Some cats like to have their stomachs rubbed. Others do not like to have their stomachs touched at all and, despite the apparent invitation, will immediately grab your hand with their paws, claws extended if they have them, and try to bite you. If this should happen, don't do the instinctive thing and pull away suddenly, or you will inflict a lot of pain on yourself. Leave your hand near the cat's body and slowly withdraw it sideways. You will quickly learn if your cat has a hands-off attitude toward her abdomen, and should warn visitors about it.

An affectionate cat that wants your attention or some petting often uses a front paw to tap you or pull at you. But an annoyed cat that wants you to stop doing something may whap at you with her paw, as if to say "Leave me alone."

Cats are not normally social with other cats, although cats that live in the same household may become close companions. When a friendly cat first meets another friendly cat in her territory, she will perform a greeting ritual. With her tail up in the air, she will rub her cheeks against the other cat's body and face to share scents. She may then lick the other cat's face. Cats that live in the same household often go through this ritual at least once a day. It is a form of reassurance and also a way of establishing dominance (the licker is submissive to the receiver of licks). When two cats live together, their dominance and submission roles may change over time. For example, if the dominant cat becomes old or ill, the other cat may take over her role.

The Body Language of Cats

	BODY	EARS	EYES	MOUTH	TAIL	VOCALIZ-ATION	WHISKERS
AFFECTION	Relaxed, slightly hunched; rubbing, bunting; tiptoes	Perked forward	Open	Closed	Relaxed up, or down	Murmur, purr, chirp	Fanned out to the side
AGGRESSION	Head lowered	Rotated backward	Slitted	Lips drawn back	Lowered, swishing	Hiss, shriek, moan, growl	Pulled back tightly against face
ANGER/ ANNOYANCE	Upright	Flattened backward	Narrow; pupils dilated	Lips drawn back	Swishing	Low moan	Pulled back tightly against face
DEFENSE	Crouched	Laid back	Open	Slightly open	Swishing	Spit, hiss	Pulled back tightly against face
EXCITEMENT	Tense, upright	Perked	Wide open	Slightly open	Twitching	Chatter or soft moan	Fanned out to the side
FEAR	Back hunched, fur puffed;	Pulled back	Wide open	Normal	Puffed	Hiss, growl	Spread out toward front
	or crouched	Flattened backward	Slitted	Normal; drooling	Down, between legs	Spit	Flattened
PLAYFULNESS	May bow down, extend paw; may pull at you; tiptoes	Perked	Wide open	Normal	May wag	Soft meow, chirp, murmur	Fanned out to the side

Cats that live in the same household often become friendly and indulge in mutual grooming.

What Stance Says

It is pretty obvious when a cat is perfectly calm. Her back will be straight and her feet will be flat on the floor.

If she is delighted and happy to see you, she may hunch her back and stand on tiptoe. This is often accompanied by head-butting; rubbing against your legs, hands, or whatever part of you is accessible; and murmuring or chirping.

When a cat is really frightened, her back may be exaggeratedly hunched up, like the Halloween cat; or she

A nervous or cautious cat.

may crouch down as low as possible. These are two distinctly different reactions to fear.

An aggressive cat will have a lowered head and somewhat curved back.

What the Tail Says

Normally, a cat's tail is held in a relaxed way, slightly curved at the tip, and hanging down or straight up and waving in the air. When she is interested or excited, her tail may twitch slightly. But her tail may also twitch if she is annoyed.

An aggressive, annoyed, or angry cat holds her tail down, close to her body, sometimes curled slightly at the tip. Her tail will swish from side to side. This is a clear warning signal.

A normal, relaxed stance.

If a cat is frightened, her tail will be down, but it may be puffed up, Halloween-fashion. Alternately, a frightened cat may lower her puffed tail between her back legs.

A very happy cat may wag her tail from side to side in a gentle manner that is completely different from the swishing or twitching tail of an angry or annoyed animal.

What Fur Says

A relaxed, healthy cat has a sleek, well-groomed coat. If a cat's coat suddenly looks ragged and unkempt, it is a sure sign that she is not grooming herself and is not feeling well. Waste no time taking her to the veterinarian!

A frightened cat's fur may stand on end all over her body. As I mentioned in Chapter Two, this is a unique feline ability known as piloerection.

Stress of any kind will cause a cat to shed excessively. Some cats react to any small change in their environment by shedding a lot. And almost every cat owner has had the experience of opening the carrying case in the veterinarian's office to be greeted by a cloud of flying fur. There is nothing much to do about this kind of shedding. It occurs even if the cat has just been well groomed.

VOCALIZATION

Cats are not as vocal as dogs, in general, with the notable exceptions of Siamese, Abys, and some Orientals. Cats have individual voices, just as people do. An owner of several cats will always recognize which one of his pets is "talking" to him.

The normal volume of a cat's voice varies a great deal. Some longhaired cats have almost silent voices, or no voices at all (perhaps a faint peep). You often have to look at them and see their mouths open to even know that they are meowing. Other cats, notably Siamese, Abys, and Orientals, have amazingly loud, strident voices. Most cats have voices somewhere in between these extremes.

Cats also differ in the amount of vocalization they

• Cat Talk •

CATERWAULING (YOWLING)	A cat may caterwaul (yowl loudly in an extended call) when she is lost or confused, in need of immediate attention, or very hungry.
CHATTERING	A cat will chatter her teeth and move her jaws up and down rapidly in a machine-gun fashion when she sees prey but is unable to get to it, as through a closed window.
GROWLING	A low, harsh, drawn-out sound of anger.
HISSING	A cat hisses when angry and usually when on the defensive. She hisses with her mouth slightly open, teeth showing.
MEOWING	This is a cat's way of "talking" and is usually used to greet friendly people and other animals. It is somewhat long and drawn out and sounds like "eee-ow." It may sound plaintive.
MOANING	A cat gives a long, low moan before she throws up a hairball.
MURMURING	Similar to a purr but less expansive. A cat often murmurs to her owner in greeting, or as a request for food or attention. A murmur can also be a thank you. It is a soft, brief, buzzing sound.
PURRING	Purring is not really a vocalization, but it is usually perceived as one. It is a soft, rhythmic, prolonged, buzzing sound. It occurs in social situations and is usually the sound a cat makes when she is happy and content. But a cat may also purr when she doesn't feel well or is in pain. In the latter cases, it shows submission (a need for help).
SPITTING	Spitting usually occurs just before or after a hiss and is a further sign of anger or annoyance. It is an explosive sound (often moist).
SHRIEKING	Cats shriek when they fight or if they are in severe pain.

regularly indulge in. Some (again, Siamese, Abys, and Orientals in particular) talk all the time, even when there is seemingly nothing to talk about. They call for attention at all hours of the day and night. It is not unusual for a vocal cat to stand in the middle of a room, open her mouth, and literally yell. This can be very disconcerting for an inexperienced owner or the first-time owner of a very demanding cat. It doesn't usually mean that anything is wrong. It just means that the cat is feeling lonely and wants attention *now*.

Most cats talk softly to greet you. They talk if they are feeling affectionate, content, or hungry. The silent types hardly vocalize at all, unless they *really* want or need something. But even the quietest cat will vocalize if she needs help: For instance, if a cat is closed in a closet or cupboard and needs "out," she will call. But if she has a very soft voice, she may be hard to hear.

It is not difficult for even the most inexperienced cat owner to recognize a happy vocalization versus an angry or fearful one.

Happy Talk

A normal *meow*, whether soft or loud, is clearly a social vocalization. It may mean "Hello," "Here I am," or "Time for dinner." It is a slightly drawn-out sound. Cats cannot sound an "m," so what we call a meow usually comes out sounding like "eee-ow." Even though it often has a plaintive note, a meow is not an unhappy sound.

Purring is very easy to recognize. It is also a social sound, made only in the presence of people or other animals. It is associated with contentment but may signify submission or a reversion to kittenhood. Nursing kittens purr loudly while they knead their mother and drink her milk. Purring is a soft, rhythmic, buzzing sound made with a closed mouth. It seems to come from a cat's throat. There are several theories about the origin of the purr: It is thought to be due to contractions of the larynx and the diaphragm, or it may arise from vibrations of blood vessels in a cat's chest.

A female cat that is in heat may purr loudly as she calls out for a tom. A cat may also purr when she is anxious, hurt, or in pain.

Cats often make a low, *murmuring* sound that is similar to a purr, and is also made with a closed mouth. It is not rhythmic and doesn't last as long as a purr. This is a cheerful sound that can be a greeting, an acknowledgment, or perhaps a request for food.

A *chirp* or *chirr* is very short and sounds something like a bird's chirp ("brrr-p"). It is a sound that a mother cat uses to encourage, greet, or call her kittens. A cat may use it with her owner for the same reasons.

An angry cat.

Angry Talk

A cat's *growl* is similar to that of any other animal. It is a long, low, harsh sound that is difficult to mistake for any other.

An angry or defensive cat may also *hiss* with her mouth slightly open and her teeth showing. A hiss is also easy to recognize. It consists of a drawn-out "ssssssss." Cats usually hiss as a warning to other cats and animals they do not like or are afraid of. But if a cat is frightened or really angry, she may hiss at a human.

Spitting goes a step further than hissing and is an additional sign of anger or extreme annoyance. A cat spits explosively, spraying moisture as she does so.

When cats are in a serious fight, they often *shriek*. A cat may also shriek if she is in severe pain.

Other Talk

If a cat *caterwauls* or *yowls*, it usually means she is lost or confused and needs help or immediate attention. She may be truly lost, or simply in an unfamiliar environment and wanting to contact her owner right away. Sometimes a cat will yowl loudly when she is very hungry. A caterwaul or yowl is a loud, escalating, drawn-out call.

If a cat is sitting indoors next to a closed window and sees a bird or squirrel right outside the glass or screen, she will often *chatter* her teeth and make a small moan in excitement and frustration. Sometimes she will not actually chatter her teeth but will simply move her jaws up and down very fast and moan softly.

When a cat is about to throw up a hairball, she will emit a long, low, *moaning* sound.

HOW CAN YOU COMMUNICATE (CatSpeak) WITH YOUR CAT?

Now that you are beginning to understand what your cat is trying to say to you through facial expressions, body language, and vocalization, how can *you* communicate with your pet?

I have already talked to you about cats' acute senses. The best way for you to communicate with your cat is by using these special senses.

Touch

Patting, stroking, and massage are all wonderful ways to communicate affection to your cat. Each form of petting can be very rewarding for both of you. Patting consists of short, soft hand motions that can be applied to any part of a cat's body. Some cats like "heavy petting," others soft stroking. It will not take long to discover which your pet prefers.

Stroking is more prolonged and usually starts at the top of a cat's head and proceeds to the base of her tail.

Real cat massage is more programmed and takes into account a cat's skeletal and muscle structures. It can be given to any portion of a cat's body with various parts of your hands and arms. There are books and videos that give detailed instructions about cat massage.

Most cats' favorite petting spots are anywhere on the head, including the cheeks, behind the ears, under the chin, around the neck, and the top of the nose. They also like to be stroked along the spine. Many cats dislike being touched on their feet and legs and at the base of their tails (especially males). And keep in mind that some cats *really* do not like to have their stomachs touched.

Hearing

Because of cats' highly developed sense of hearing, they can easily detect differences in your tone of voice. When you want to show affection for your cat, talk in a soft, crooning voice and she will respond immediately. Of course, if you are annoyed with her and want her to stop doing something, a sharp vocal reprimand will usually get her attention.

Cats also respond in a positive way to normal conversation. A cat that is fond of you will enjoy being talked to regularly as you go about your normal routines. "I'm going upstairs now, Misty," or "Now I think I'll sit down and read," and so on. The more you talk to your cat conversationally, the more accustomed she will become to your voice and bond with you. She will then become attentive to your moods and your wishes.

Chapter Eight

How to Become Top Cat

To become Top Cat in your household, you must understand your cat's body language and vocal communications, both by observing her and by reading the preceding chapters in which I have outlined basic cat behaviors and abilities. In other words, you have to be able to interpret the signals your cat is giving to you when she performs various actions. Once you have mastered a degree of comfortable communication with your pet, you will be ready to go on and begin to teach her what you want.

First establish a relationship with your cat that makes

it perfectly clear that you are her sole and all-important caregiver. To do this, provide her with a comfortable home, food on a regular basis, and fresh drinking water. You pet, massage, and groom her regularly (as I have described in Chapter Six), so that she feels good and becomes close to you. You play, interact, and talk with her as often as possible so that she becomes used to the sound of your voice and further bonds with you. You provide her with stimulating toys and "furniture," such as a sturdy scratching post, for exercise and diversion. Most important, you see to it that her bodily functions can be performed in comfort (in other words, you keep her litter tray clean, fresh, and always accessible). If you do all of these things, your cat will soon learn that she needs you and your affection and approval, in order to remain comfortable, happy, and well cared for, and she will want to please you.

Establish a Routine

In general, cats are creatures of habit, although less so than dogs. Their habits sometimes change for reasons that are mysterious to us. For instance, a cat may suddenly change her usual napping location for another completely different spot. But, just like any animal, a cat will appreciate knowing what time(s) of day she will be fed. She will also respond to grooming, petting, massage, and play on a regular basis, more or less. This doesn't mean you must stick to a rigid routine, but it does mean that if you normally give your pet her dinner at six o'clock, she will not be happy if you delay her feeding until ten (among other things, she'll get very hungry!). Set your own timetable. If you are always away during the day, give your cat attention and playtime in the evening. If you are at home during the day, groom and play with her after breakfast or lunch. It doesn't matter when you opt to interact with your cat, as long as you do so on a fairly consistent basis, at approximately the same time, so that she knows what to expect and when.

Cats differ from dogs in a very important way. They are not pack animals, used to being always surrounded by loving family members. They are loners, so they do not normally suffer from separation anxiety. Although they may miss you and the stimulation and love you provide and will greet you happily when you return home, they do not usually become terribly upset and indulge in destructive behavior when you are not home. They normally sleep most of the time you're out. (Of course, there are always exceptions. A very dependent or very social cat may act out in inappropriate ways if she is left alone for long periods of time. Some cats even act out when being cared for by a sitter who comes in once or twice a day to feed them.)

A very important routine that you will need to establish early on is bedtime—yours. Because cats are normally nocturnal and sleep most of the day, they naturally want to play and interact with you at night. This does not fit in with most people's routines. It is no fun at all to be awakened in the middle of the night by a cat running wildly over your face, or by a loud call to come and play. Another favorite middle-of-the-night game is to knock everything off the top of a dresser or night table in your bedroom. This is guaranteed to wake you up with a start, as your key chain, wallet, coins, and heaven knows what else plop loudly to the floor.

You must give the message right away that this behavior is not acceptable, with kittens in particular. Give your cat some extra attention just before your bedtime and make sure she has something to eat (or leave out a dry snack for her to nibble on during the night), *and* totally ignore her nighttime advances. She will usually quickly get the message that, for you, the night is for sleeping, not playing. If her midnight antics become a persistent, uncontrollable habit, see Chapter Nine, for other ways to deal with it.

Take Control

There may be other things you want to control in addition to teaching your cat that you do not want to interact with her while you are sleeping. If you read your cat's signals (body language and vocalization—see Chapter Seven) and anticipate problems through knowledge, you may be able to prevent certain behaviors before they become real problems.

For example, all cats love to jump and climb. If you see your cat coiled and ready to jump up on the étagère that is filled with fragile glassware, act quickly, before she even has a chance to jump. Immediately clap your hands and loudly say "NO." Or spritz her with water. If you are firm enough, she may never even *think* of jumping up there in the future. (Again, if this behavior becomes a persistent problem, see the next chapter for some solutions.)

Some owners want to teach their cats to stay off surfaces such as kitchen counters or particular pieces of furniture. The same actions will often help offset the problem before it becomes a habit. Obviously, if you are away from home for long periods of time during the day or evening, you will have to either cover the furniture or close the cat out of the rooms where you do not want her to trespass until she is well trained, or conditioned, not to do so. Cats are notoriously curious and, given a chance when nobody's around, will explore any forbidden area.

The same solution works for furniture scratching. Try to catch your cat *before* she lays a paw on the velvet sofa, while she is just thinking about it. Clap your hands, say "NO!" and quickly place her paws on her scratching post. Praise her lavishly if she scratches the post. If you learn to read your cat's body language and get a jump on her actions, you can often stop her from inappropriate behaviors before they become a problem.

Annoyance or aggression is another natural, self-protecting, but unacceptable response in a cat. In this case, her body language is usually clear and easy to read. If your cat suddenly lays back her ears and emits a low growling noise when you are petting or grooming her, it is a clear sign she does not like whatever you are doing and you had better back off. If you immediately stop what you are doing, you will avert a nasty scratch or bite. If you ignore her warning signals, however, you will be sorry. What's more, you will have allowed a very antagonistic feeling to develop on both of your parts. If you allow (or, in a sense, encourage) your cat to get away with an angry response, you will have diminished your control and your mutual sense of trust. If she does strike out at you, immediately say "NO!" and clap your hands. Do not hit her, as this will only increase her anger, but you can tap or flick the tip of her nose with your finger. This is a disciplinary action a mother cat will often take with her paw when a kitten is misbehaving, and it usually works. See Chapter Nine for more solutions to common cat problems.

BASIC CAT TRAINING

Yes, cats *can* be trained! Once you and your cat have established a comfortable routine and you have taught her to conform to most of the rules of your household, you may decide that you would like to go a step (or several steps) further. If you really want to become Top Cat and create a bond so that your pet will listen and respond to you, you will need to teach your cat certain commands, some very simple and others more complicated, depending on how far you really want to go.

Of course, just as when you are establishing a routine and taking control, you and your cat must have a mutual understanding before undertaking a training session. You need to learn how to communicate with your cat and understand what she is saying to you. Then you will be able to control her, train her, and understand her behavior problems (see Chapter Nine).

As I mentioned before: With dogs, you teach them to respond to commands. With cats, you ask them to do things. Just as a mother cat teaches her kittens how to hunt by demonstration (bringing home a mouse) and creates a learning environment, I work with cats' natural instincts to train them.

When training a cat, keep it simple. Teach her one thing and reinforce it, then review it each time you add something new. Make training a comfortable, relaxed process. Work in short sessions, and stop as soon as the cat's attention wanders (usually after no more than three to five minutes). If possible, repeat each lesson two or three times a day.

For cat training, I cannot stress enough the three P's—Patience, Persistence, and Praise. These three "golden rules" are especially important with cats, because they have short attention spans and very little patience. If you want to train your cat, you will have to bring along a lot of your own patience. It can be a rewarding and satisfying experience to really become Top Cat and train your cat to respond to your wishes and commands. Not every cat is equally responsive—if you have learned to read your own pet's language, you will be better able to work with her on her level.

Remember that the entire process should be a relaxed, enjoyable time for both of you. Your cat should learn to enjoy her training lessons, and in the end, you will have created a more responsive, pleasing pet.

At the end of each training session, put the cat in her carrying case, carry her to another room, open the case, praise her, and give her a treat and a pat. Do this on a regular basis, and soon your cat will learn to think of her carrier as a sanctuary; a nice, safe place to go and relax. Eventually she will learn to go into the carrying case on command (for example, "Go in your case").

Training Equipment

The best location for training a cat is a waist-high table with a nonskid rubber mat on top. A bathtub mat, shower mat, or rubber-backed doormat is best, because it will prevent the cat from slipping. Professional trainers often use special carpet-covered tables. Do not simply cover a table with a towel; it will slip and make the cat feel very insecure. This equipment is especially important if you want to progress to serious training. The table should be placed in a corner so that it has two side walls for security and for preventing escape. If possible, do not use the same table for training that you use for grooming. The training table should be a special place just for lessons. If it is impractical for you to have a special training table, a kitchen table or countertop covered with a nonskid mat can be a satisfactory substitute.

While training the cat, you should stand approximately twelve to fifteen inches away from the edge of the table or counter. (This is why I recommend a long stick or dowel for treats.)

You should also have a toy to help the cat focus. A stick with a feather "bird" on a string is particularly useful. (Note: Be sure to always put this toy safely away so the cat cannot become tangled in the string when you're not around—see my mention on the danger of linear objects in Chapter Five.)

You will need a two-foot stick or dowel on which you can place a treat for the cat—a wooden cooking spoon or spatula works well.

Have your cat's carrying case on hand. Part of her basic training will be to learn to go into the carrier on command.

A soft fabric or leather harness or collar and a leash are necessary. *Never* use a metal collar or harness on a cat.

Training Inducements

Your voice is the best inducement in training your cat. Praise and encouraging noises are a must as you go along. You can also use mouth noises such as clicks and kissing sounds to further help your cat focus on you.

A food treat works well for most cats as a reinforcement for the vocal commands you are teaching. When she responds to a vocal command, a food treat can be a reward in addition to praise. However, you have to be somewhat careful not to overdo, especially if you have a "pig" cat, or she will quickly become overweight.

Petting, stroking, and massage are wonderful incentives for a cat. There are very few felines that do not respond in a positive way to gentle handling.

Training Equipment and Inducements

Equipment

- Waist-high table topped with a nonskid rubber mat; the table should preferably be placed in a corner so that only two sides are open, for better control
- Stick with a string and feather or other light, dangling object, to get a cat to focus
- Two- to three-foot stick with a spoon or indentation at end for treats
- Sturdy carrying case
- Harness or soft collar and leash

Inducements

- The sounds of your voice—words, praise, clicking noises, "tch-tch," kissing noises—for focus
- "Feel good" rewards: petting, stroking, massage
- Food treats

Teach Your Cat Her Name

The first step in training a cat is to teach her to recognize her name and respond to it. Always use a cat's name when talking to her. For example, "Here's your supper, Misty,"

or "Hi, Misty, would you like to be combed?" When she walks into a room, say "Hello, Misty," and so forth. A cat will respond to the sound of her name and the way you say it. Always use a cat's name in a friendly, affectionate way. Do not use a cat's name when you are scolding her, or she will associate it with a negative feeling.

To begin to teach a cat her name, call her by name. You may initially need to use some encouraging mouth noises to get her to come, especially if she is a young kitten. Shake a treat box, or open a can of food and call her. This may do the trick. When she responds, praise her, pet her, and give her a treat. Use her name again in a positive way: "Good, Misty. Nice, Misty." When she comes, make it into a love session. Massage her a bit, lavish her with praise, and give her a treat so that she associates coming on cue with a positive experience.

Teaching a cat to come when called.

As your cat begins to respond to her name, try calling her from other rooms in the house. If she comes, make a really big fuss so that she begins to associate coming when called with a *very* positive experience.

You can embellish the training element of coming when called. Once your cat is responding to her name on a regular basis, put her into her carrying case when she comes to you. Take her to a different location in the house, let her out, praise her, and give her a treat. This is part of my *basic training* to accustom a cat to accept a carrying case and eagerly anticipate a trip in a carrier so that she will hop into it on command. Because any cat will have to be taken places in a carrying case many times in her life, it is really nice if she can learn to welcome it as a positive experience and not a punishment to be avoided at all costs! (If the carrying case only represents a trip to the veterinarian to have shots, who can blame her for trying to avoid it?)

Teach Your Cat to Sit on Command

Place the cat on the table. Gently place one hand over her body and run your hand along her spine from head to tail. Push down on her rump as you say "Sit." You can apply more gentle pressure if it is needed to get the cat's rear end down onto the table. Once she is in a sitting position, praise her—"Good, Pearl"—reward her, and let her go. Go through this routine several times until your cat responds to the command and will remain sitting while you move farther away from the table.

If she gives you a hard time and doesn't want to sit, press down on her back and leave your hand in place. Praise her, but do not give her a treat. Only give her a treat when she sits on command, without hand pressure.

Teach Your Cat to Stay

This command requires a lot of Patience, Persistence, and Praise, but the end results will be well worth your efforts. A cat that has learned to sit and stay will behave perfectly in any situation, even at the veterinarian's office, and will no longer require unpleasant restraints. This also serves as an important safety measure: A cat that has learned to stay can be stopped from running out of an open door or window.

Basically, this is a long sit. Begin with the sit command. Once the cat is sitting, place your open hand in front of her face with your palm toward her face. Say "Stay," slowly remove your hand, and pause for a few moments. If she stays quietly, praise her and give her a treat. If she breaks out of the stay, immediately put her back into a sit and repeat the process. Step back farther and continue to say "Stay." Have her stay longer each time, then gradually back away from the table. Ideally, you will be able to have your cat stay when you are six feet away from her, wherever she is.

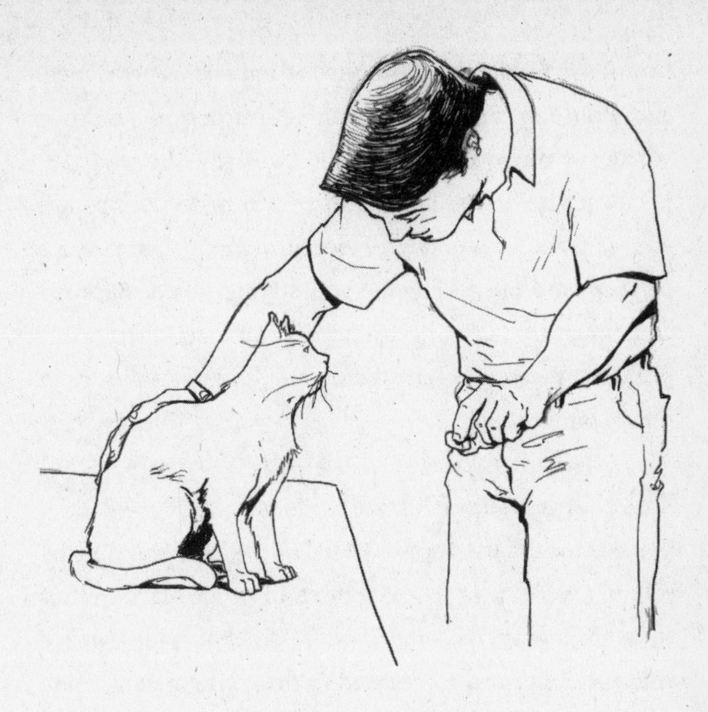

Training a cat to sit-stay.

Teach Your Cat to Lie Down

Review the sit command. Begin with the cat sitting on the table. Then put your left hand on the cat's shoulder and your right hand behind her front legs just above her ankles (you may change hands if it's more comfortable). Say "Down" and slide the cat's front legs forward while pushing down on her shoulders. Keep the pressure light and continue saying "Down" or "Lie down." Hold the position, pause, say "Good, Grace," and give her a treat.

While the cat is lying down, massage her shoulders

so that she feels good. With a lot of Patience, Persistence, and Praise, your cat will eventually learn to lie down whenever you touch her shoulders and say "Down."

When your cat has mastered the down command, you can add down-stay. Proceed as above, with your palm toward her face while she is lying down. Back off and, when she stays, praise her and give her a treat. Repeat this as often as necessary until your cat will stay in the down position.

Take It a Step Further

Once your cat has mastered sit, down, and stay on the training table, it is time to move to other places, other rooms. Change the environment so that your cat becomes conditioned to respond to these commands wherever she is. If she doesn't respond right away in a new location, take her back to the table and go through the routine again. If she still doesn't respond, try going through the routine wherever you are. Patience!

Teach Your Cat to Walk on a Harness (or Collar) and Leash

This is something that not all cat owners will want to do. But if you are ever going to travel with your pet or take her to a new environment, such as someone else's home, you will have better control of your pet and she will be much safer if she is wearing a harness or collar and leash,

A cat wearing a harness and leash.

which will keep her in focus and prevent her from indulging in mindless flight behavior and into possible danger. (Even a cat that has learned to stay on command may fail to focus and might panic in certain situations.) The leash is a tool that will act as positive behavior modification, otherwise known as behavioral engineering.

A harness works better than a collar for most cats. It provides better security and control, and the cat cannot slip out of it. But some cats go berserk when harnessed and should wear a collar instead. The collar should not be too tight, but it needs to be comfortably snug so that the cat cannot pull her head out of it. To a certain extent, your cat's body type will determine whether a harness or collar works best. A cat with a long, thin neck and slim body—such as an Abyssinian—will probably be more secure in a harness that she cannot slip out of. On the other hand, a cat with a short, thick neck, or one with a lot of fur around her neck and a chunky body—such as a Persian—will be more comfortable in a collar.

Put the harness or collar on the cat and let her walk around the house in short sessions with it on, for as long as it takes until she seems comfortable. Then attach a leash to the harness or collar and let the cat drag it around for a few minutes. Increase the time each day, again until she is at ease dragging the leash. Be sure to keep her in sight while the leash is on so that it won't snag on something and trap her—this can really frighten a cat!

When she is able to walk around in a normal way when dragging the leash, pick up the leash and follow her around the house. Do this on a daily basis for five to ten minutes and your cat will soon become used to it. As you walk along with her, pick up the cat and walk a few steps holding her, then put her down again. Eventually your cat will learn that this is a comforting ritual and will feel secure when walking with you holding the leash. You can use a helper to hold a treat in front of the cat to encourage her to walk with you.

A cat that is well leash-trained can go into any situation with her owner and feel secure and safe. She will even learn to ignore major distractions such as other (strange) cats in the room or loud noises, and she will remain perfectly calm when approached by people who are new to her.

SOME SIMPLE TRICKS

Once your cat has learned to walk on a leash and respond to the commands above, she is ready for some simple tricks. I will discuss some more unusual tricks in Chapter Ten, but for now I would like to teach you how to train your cat to shake hands and sit up and beg.

Teach Your Cat to Shake Hands

This is a trick everyone will like—it makes a cat seem very smart, "just like a dog." It also encourages your cat to communicate with you. All cats use their front paws a great deal. For instance, when a cat wants attention or petting, she will often pull gently at her owner's hand or arm with a front paw. This trick is a natural outgrowth of that action.

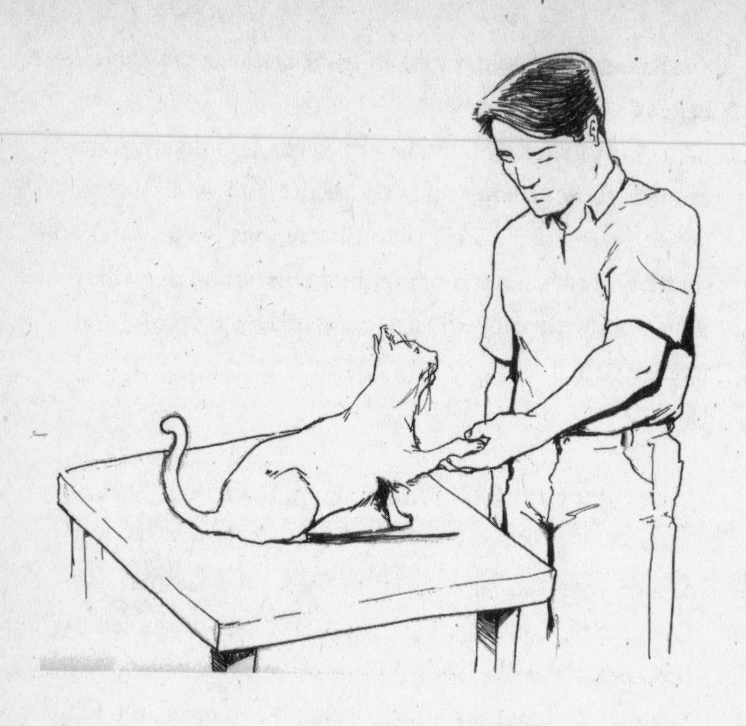

Teaching a cat to shake on command.

Begin with the cat in a sit position on the training table. Extend your right hand as if you are going to shake hands, then nudge the cat's right leg behind her elbow as you say "Shake." The cat may immediately raise her paw. Hold her paw gently, praise her, and quickly give her a reward. Do this over and over until you no longer have to touch her leg for her to respond to the command. She will have become conditioned to shake on command.

As with the previous training routines, once your cat becomes proficient at shaking hands on the table, move her to different locations so that she learns to shake wherever she is.

Teach Your Cat to Sit Up and Beg

This is another action that is associated with dogs, and many cat owners enjoy proving to their friends that cats can do it, too.

Training a cat to sit up.

Start with the cat sitting on the training table, facing you. With this trick, it is a good idea to have the cat on a leash so that you can exert a gentle pulling motion to encourage your cat to sit up. Stand a short distance away and place a treat on a wooden dowel or stick. Hold the stick a few inches above the cat's head. The cat will focus on the stick and reach up with a front paw to try to reach the treat. As she reaches up, say "Sit up" or "Beg" and pull the stick higher above her head. The cat will then reach up with both front paws. After a couple of seconds, say "Good, Martha," give her the treat, and praise her lavishly. After she has mastered this, have her stay in the begging position longer, walk away a foot or two, and then reward her.

Once your cat has learned to sit up and beg on the table, transfer her to the floor and repeat the process so that she learns to beg on command, wherever she is.

For more tricks, see Chapter Ten.

Keep It Happy

In any training process, both you and your cat should always end up with a positive feeling.

Always praise her and give her some affection after every training session, even when it has not been very successful. (Don't, however, give her a treat when she has not responded in an appropriate way.) Remember:

Patience, Persistence, and Praise—the all-important keys to any training process.

ARE YOU TOP CAT YET?

That's for you and your cat to determine. If you have established a good daily routine with your pet and can communicate your wishes to her, you are well on the way to becoming Top Cat.

When you have mastered the steps in basic cat training, you are definitely on the way to being the Top Cat in your household. Should you choose to go on to some simple tricks, there is no question about it: You are definitely Top Cat!

What's more, your cat will now be prepared to take the Cat Fanciers' Association Feline Good Citizen test (see the Appendix) and pass with flying colors!

Chapter Nine

Some Simple Solutions to Common Cat Behavior Problems

No matter how well you may understand your cat and be able to work with her in general, she still may develop a problem behavior. By "problem behavior," I mean something that is destructive, annoying, or dangerous.

The seven most common kinds of problem behavior are: aggression toward people or other pets; destructive behavior, including clawing furniture and other household items; dumping garbage; hyperactivity; plant eating and digging in soil; house soiling; and walking or sitting on prohibited surfaces. Many of these behaviors can take several different forms.

I am a firm believer in *prevention*. If you use my formula of the Three P's—Patience, Persistence, and Praise—you may be able to modify these behaviors before they become problems.

One of the most important steps in preventing several kinds of behavior problems is having your cat neutered at the appropriate time. Not only is spaying (for females) or castrating (for males) an important step in maintaining a cat's health throughout her or his lifetime, but it will also have a definite positive effect on a cat's behavior. It will usually eliminate or cut back on aggression toward other cats or humans. A neutered cat, especially a male, is much less apt to indulge in marking behavior that leads to house soiling. Neutering will also result in less hyperactivity and frantic, seemingly mindless wildness. A neutered cat will be a calmer, more responsive pet.

Neutered or not, the sexes still retain some of their inherent, instinctive behavior characteristics. Male cats tend to care more about supremacy and strive to become the "boss cat" in a household. Females are generally better hunters (to feed their kittens) and have a stronger sense of territory.

A Few Handy Problem-Solving Tools

- A spray water bottle
- A shake can (empty can filled with small stones or pennies)
- Two-sided sticky tape or sheets of Con-Tact paper
- Mothballs
- Aluminum foil
- Bitter-tasting spray
- A roomy crate (Although cats hate to be confined, sometimes a crate, used for a short duration, is a good solution to a problem.)
- Catnip or aerosol can of catnip oil (Note: Dried catnip stays fresh and aromatic in the refrigerator.)

HOUSE SOILING

House soiling is the most common problem I see for cat owners. Either the cat soils at random all over the house or regularly soils in one particular area outside of the litter box. Cats often like to use a bathtub as a litter box; others prefer the middle of their owner's bed.

The very first thing an owner should do if a cat is regularly avoiding her litter box is to have her checked out by the veterinarian for a possible health condition that may be the cause of the problem. Once a cat is given a clean bill of health, you can be sure the problem is behavioral, not medical. Here is an example that proves this point.

Case History

Mia, a seal-point Siamese who had always used her litter box, suddenly began wetting on her owner's bed. The urine had an odd, pungent smell. I suggested that the owner take Mia to the veterinarian. It turned out that the cat had a urinary infection, and as soon as she was put on medication and a proper diet, the problem was solved. This, then, was not a behavior problem.

Litter Boxes

No self-respecting cat will use a litter box that is soiled or smelly (remember that your cat's sense of smell is many, many times more acute than yours). In the case of an unacceptable litter box, a cat will often eliminate *near* the box, just not *in* it. She is sending you a clear message that she is trying to do the right thing but just cannot bring herself to actually go into the dirty box. If this is the case, the solution is very easy. Routinely remove all solid waste as soon as possible (a slotted scoop is all you need), add

A well-mannered cat eliminates in her litter box.

some fresh litter if the box seems wet, and change all the litter at least once a week. If you have more than one cat, be sure to provide enough litter boxes. (See Chapter Five for more about litter boxes.)

Another common reason for a cat to disdain a litter box is its location. Cats prefer some privacy when they go to the bathroom. Some appreciate a small folding screen around two sides of a litter tray, for seclusion. They do not like to be where there is a lot of foot traffic and activity, such as a kitchen or playroom. Also, because they feel especially vulnerable when they are squatting, they like to be able to look around. The ideal spot for a litter box is in a small room such as a bathroom, or in a basement (if it's in the basement just be sure that you don't forget to keep it clean). A corner location is ideal; the cat is protected on two sides so no one can sneak up on her from behind.

Litter

Cats have a natural self-protective instinct to bury their waste, so we must provide our house cats with the appropriate material to eliminate in (outdoors, they will use soft earth).

There are many types of litter on the market, and some cats will develop a definite preference that sometimes is so strong they won't use any other type of litter. For example, declawed cats often do not like clay litter because it is rough and hurts their feet; other cats *prefer* the harder clay litter to kick around. If you have more than one cat, you may need to provide a different kind of litter for each cat. When you adopt a kitten or cat it is always a good idea to find out, if possible, what type of litter she is used to so you can continue with the same kind. A simple solution to many elimination problems is to change the type of litter you provide. Let me give you an illustration:

Case History

Blue, a beautiful blue Persian, was using his owner's entire house as a litter box. His owner did not know what the matter was, but she could not live with the constant mess and was on the verge of giving Blue away when she called me in for help.

I found the clay litter in his box was perfectly clean and there were no problems with the box itself. But I also knew that longhaired cats often do not like clay litter because it sticks to their fur. I suggested a simple solution—change the litter.

We worked out a plan in which the owner got a second litter box. She filled each box with a different kind of litter: one with sandy, soft, clumping litter; the other with cedar-based litter. In less than a week, Blue discovered that he loved *cedar-based litter, and the problem was completely solved.*

Marking Behavior

Sometimes eliminating outside of the litter box is due to marking behavior. Even neutered males and either spayed or unspayed females may indulge in marking. This often happens when there are several cats in a household, or when a newcomer is brought into a house where older cats already live.

Case History

Ike and Mike were eight-year-old neutered male tabbies that lived happily in the same household and shared two litter pans. One day their owner brought Molly home. She was a delightful calico kitten. Ike and Mike seemed to accept her calmly and even happily, but within days they were spraying urine all over the house. Their owner called me, frantic.

I immediately recognized the problem. Ike and Mike were indulging in marking behavior, clearly and unmistakably showing Molly that this was their *house and* their *furniture!*

The solution? I told the owner to put Molly in a small room with a bed, water and food, and her own litter pan. (Note: A crate large enough to hold a litter pan and food and water bowls would also work.) Although I know that cats very much dislike confined spaces, it was the only solution in this case, and Molly's "incarceration" would hopefully not last very long.

As soon as Ike and Mike realized their territory was no longer threatened, they went back to their usual gentlemanly habit of using the litter pans. I told the owner to gradually begin to let one or the other of the "boys" into Molly's room for a short visit. Each time Ike or Mike behaved in a friendly way toward Molly, the owner praised and petted him. After about a week or two, as the two males became used to Molly and realized she wasn't really a threat to them, they allowed her the run of the house and even let her use their litter boxes. It just took a lot of Patience, Persistence, and Praise on the owner's part to introduce the new cat gradually.

DESTRUCTIVE BEHAVIOR

The most common kind of destructive cat behavior is clawing furniture, rugs, curtains, or other household items. Along with house soiling, this behavior problem comes high on many a cat owner's list. Some cats also indulge in other types of destructive behavior, such as knocking things over, pulling books and bric-a-brac off shelves, chewing things, and so forth. I'll talk about clawing first.

Clawing

As I mentioned in Chapter Five, a sturdy scratching post is a *must* for any cat, even one that has been declawed. From her first day in her new home, a kitten or cat should be conditioned to use her scratching post—and only her scratching post—for clawing. If you can accomplish this right away, it will prevent clawing problems from ever developing.

Always keep a cat's nails well trimmed. This will cut back on one reason for clawing behavior.

To teach a kitten to claw on a scratching post (or an adult cat that has not yet learned to use one), do the following:

- Put the post in a location where it stands firm and cannot topple over. Many people put it at the corner of a piece of furniture (a likely spot for a cat to begin clawing) to redirect temptation.
- Rub some catnip, or spray catnip oil, into the surface of the post. You can also hang a dangling toy from the top of the post.
- Encourage the cat by placing her front paws against the post while you stroke her and praise her. Move her paws in an up-and-down scratching motion.
- When she scratches the post, praise her lavishly.
- Make using the scratching post enjoyable for your cat. Turn it into an even better playtime with a dangling toy. Praise her and pet her when she uses the post.

These steps work well for the majority of cats, but there are always some that simply do not seem to "get it." The problem may be the surface of the scratching post. Most cats prefer a rough surface, such as rope or bark (note that bark is very messy, as it sloughs off), but some may like a different type of surface. It is worth trying different kinds of posts to protect your furniture.

If your cat is one of those individuals that continue to claw in places other than her scratching post, you will have to resort to stronger methods. If you are at home and hear the telltale ripping of claws into fabric, immediately rush into the room, clap your hands, and say "NO!" Alternatively, spritz the cat with water from a spray bottle, or shake a can full of pebbles or pennies, any of which will startle the cat and make her stop what she is doing. Then, if she is still in the room, and when she has calmed down, put her paws against the scratching post, then pet and praise her.

If you are away from home a lot—and this is usually when the damage occurs—cover your furniture with plastic. This helps, because most cats dislike the smell and feel of plastic. Blown-up balloons placed underneath a sheet also provide a most unpleasant surprise for a cat when she places her paws on the sheet and the balloons pop.

I have discovered that a very good tool to prevent scratching (and to keep a cat from getting up on a surface where you do not want her) is sticky tape. Get some two-sided sticky tape and place it in the areas your cat scratches or jumps up on. Alternately, if it is a large surface, such as a carpet or rug, place a sheet of Con-Tact paper, sticky side up, on the surface you want to protect.

As soon as the cat's paws touch the sticky surface, she will be repelled and will pull away.

Here is how I solved one owner's clawing problem:

Case History

Mabel had a big five-year-old orange tabby named Jessie. Jessie was a love, but he had destroyed every piece of furniture in her living room. It was literally in shreds. For five years Mabel had put up with the mess, but now she wanted to re-cover her furniture and didn't know how to protect it.

When I got to her apartment, I noticed she had a scratching post that was covered with soft, carpetlike fabric. Obviously, Jessie needed something to scratch that had more resistance—the sofa and chairs, for example. He was a big, strong cat, and the soft scratching post didn't provide him with nearly enough exercise. The first thing I suggested to Mabel was that she buy a scratching post covered in something rougher. I also suggested that she get a second post to put in another room.

I told her to rub catnip, or spray catnip oil, onto the surface of the post. I then suggested that she buy two-sided tape and put it in every spot Jessie usually clawed, paying particular attention to the corners of the sofa and chairs.

It was important for her to condition Jessie to use his new scratching post(s) before *she had her furniture re-covered, so he would never lay a paw on the new upholstery. Once a cat develops a certain favorite clawing area, he very often will return to it.*

A month later, Mabel called to say that Jessie loved his scratching posts and her new furniture was unscathed.

Catnip as a Training Tool

CATNIP IS AN HERB MOST CATS REALLY LIKE BECAUSE OF AN OIL (nepetalactone) contained in the leaves. Kittens do not respond to catnip, and it holds no enticement for some adult cats.

Cats who love it react in different ways: Many roll in it, purring wildly; others lap it up and go on a "crazy binge," running around, leaping in the air, and acting happy; some may eat it and go immediately into a deep, contented sleep. It acts on cats' sensibilities much the way a couple of scotches or marijuana acts on people. The important difference is that it is safe, nonaddictive, and the effects wear off quickly.

Catnip can be a useful training tool. Rubbed into the surface of a scratching post, it can act like a magnet to entice a cat to scratch on it instead of the sofa. It can also be used as a nocal inducement instead of a treat when training a cat. Catnip toys (toy mice or other small objects filled with catnip) promote exercise in even the most sedentary felines and may become favorite items for retrieving.

Dried catnip and catnip toys are found in pet-food stores and supermarkets. Dried catnip keeps its freshness and scent when stored in a closed container in the refrigerator. (A word to the wise: Catnip grown outside the United States may contain a lot of pesticides, so check labels.) Catnip oil is also available in aerosol cans—handy for spraying on a scratching post.

Cat owners often grow catnip in a windowsill container for a cat to nibble at will. However it is offered, a catnip-loving feline will find it a definite enhancement to her life.

Other Destructive Behavior

Cats can become destructive in other ways besides clawing. Sometimes a cat will go on a rampage and literally trash a house, knocking everything on the floor, breaking and destroying objects, chewing books and cushions, and so forth. Here is a case I recently had with a destructive cat:

Case History

Joy was a delightful, extremely affectionate Abyssinian. She was devoted to her owner, Jane, following her from room to room, sleeping on her bed, and always curling up next to her whenever she sat down.

The problem occurred whenever Jane went out, even for just a few hours. When she came home, the apartment was completely destroyed—and it was getting worse! Jane called me in desperation.

It was obvious to me that Joy was suffering from separation anxiety *whenever Jane left her alone. This is a syndrome most often associated with dogs, but especially devoted cats can suffer from it, too. These cats will become completely panic-stricken when their beloved owners leave them alone. They will react mindlessly and wildly, in a manner somewhat similar to that of a child having a tantrum.*

The solution is to provide the cat with a calm experience and a secure place to be until her owner returns. I suggested to Jane that whenever she planned to go out, she should calmly pick Joy up and put her in a carrying case or small crate. When she returned home, she should praise Joy and make a fuss over her. After a few weeks, Joy could be moved to a small room, such as a bathroom, where there was little to destroy. (Note: A roomy crate will provide the cat with the same sense of security.) She could have something to sleep on, a snack, and some water, but nothing else (except perhaps a favorite toy). Before she left, Jane should praise and reassure her and stroke her gently. If Joy remains calm in the small room or crate, she can gradually be transferred to a larger room until she learns to stay quietly and calmly while her owner is away from home; then she can have the run of the house.

Jane performed all of these steps (again, with Patience, Persistence, and Praise) and now can leave Joy alone when she goes out without coming home to a disaster.

HYPERACTIVITY

Hyperactivity in cats is another very common problem. One of the most annoying types of hyperactivity is when a cat regularly wakes her owners in the middle of the night or very early morning, demanding attention. Another form is sheer wildness—climbing the curtains, racing around mindlessly, and in general creating a scene.

Hyperactivity is usually due to a high energy level and boredom, although it may have a medical basis. If it persists, seek veterinary help.

Waking Owners/Keeping Them Awake

Cats are naturally nocturnal, and some just don't seem to understand that their owners like to sleep quietly during the night and early morning. They choose those times to persistently demand attention. Take the case of Toby:

Case History

Toby was a perfectly normal gray British shorthair. During the day, he slept, played, and ate just like any other cat. His owners, Beth and David, worked at home and liked to stay up late at night and often "slept in" in the morning. That is, until Toby came into their lives. Every morning around five, Toby decided it was time for everyone to get up. He jumped up on the bed, pushed his face into one of his sleeping owner's, meowed and purred loudly, sometimes patted their faces and vigorously kneaded whatever part of them he could. If he got no response, he proceeded to gallop (or thump) up and down the hallway, into the bedroom and out, and back up onto the bed, "talking" loudly all the while. Nothing they did would quiet him down. When they ignored him, he simply escalated his actions. Bleary-eyed, they called me for help.

I thought of a couple of simple solutions. First, it was possible Toby was hungry. I suggested that, instead of feeding him his main meal at six P.M., they wait until just before their bedtime (an alternate suggestion if your cat eats dry food is to leave an ample snack out for him to nibble on during the night). This way he would be full and satisfied all through the night.

Second, it was possible he was bored. Perhaps a really active play session just before bedtime (and after dinner) would tire him out and make him more content to rest at night.

Third, I suggested that if he persisted in annoying them during the night, Beth and David each put a spray bottle of water on their bedside tables. The minute Toby began his "act," whichever one of them was closest to him should say "NO!" and spritz him with water to stop him in his tracks.

They did all three things, and within a week, Toby was able to stay quiet until they were ready to get up and give him his breakfast.

Other Forms of Hyperactivity

Sometimes kittens and cats will indulge in silly or wild "fits," running around the house, practically bouncing off walls. If this happens occasionally, it is usually not a problem; but if your cat regularly has bouts of wild behavior that end up in dangerous or inappropriate actions such as climbing the draperies or the bookcase, it may become one. If you have more than one kitten of the same age, they often encourage each other in wildness.

Most kittens will outgrow this kind of behavior. But it is obvious that a highly active kitten requires the opportunity to be busy and to be challenged. To help your kitten calm down, give her lots of intelligent games to play. A Ping-Pong ball, for example, can provide endless hours of active play, especially if you toss it for your cat.

There is also a game on the market that your cat can play alone, with a Ping-Pong ball in a circular gutter. Toys on a spring that bounce back and forth can be intriguing to any cat, as are dangling objects. (Note: Be sure to put away toys that are dangling on a string when you are not around to supervise.) In other words, give your hyperactive kitten or cat plenty to do and her energy will be challenged in appropriate ways.

AGGRESSION

Aggression in cats is a very complex subject. Cats may be aggressive toward other cats in the household, toward dogs, and toward humans, or only one human in particular. It can be a very troubling manifestation that is difficult for an owner to deal with.

Stress may be a contributing factor. Cats vary considerably in the way they handle stress. Some cats seem to be able to roll with the punches no matter what happens, but others may react with aggression. If there are a number of cats in a household, or if there is simply a lot of very noisy, chaotic activity all the time, it can lead to stress in a cat. A change in the household can also stress a cat. For instance, if a beloved family member goes away, or if a new person comes to live in the house, it can promote stress in some cats.

If a kitten is taken away from the litter when she is too young, she will not have been properly taught by her littermates and her mother how to control her aggressive feelings. This kitten will be very difficult to cure of overt aggression.

There are a number of types of aggression in cats. I will discuss only a few of the types most often encountered with pet cats and make some suggestions about how to deal with them.

Petting-Induced Aggression

This is a form of aggression that is often very puzzling to owners. A cat seemingly encourages petting, then may suddenly attack your hand, clawing and biting viciously and painfully. Frequently, a cat will roll over on her back and seem to invite you to rub her stomach. But as soon as you touch her tummy, she will grab your hand and attack with teeth and claws. Male cats, especially those that have not been neutered, may react the same way if you rub them at the base of their tails.

To keep this from happening again, do not give in to requests for that particular form of petting, and warn visitors about this habit of your cat. This seems to be a form of *dominance aggression* that is triggered when you touch certain areas of a cat's body. It can often be cured by a training program such as the one I outlined in the previous chapter, when you become Top Cat.

Typical stalking posture.

Meanwhile, to avoid damage to your hand, do not pull away instinctively; simply relax your hand and slide it out sideways, between the cat's paws, while you say "NO" firmly.

Predatory Aggression

Cats are naturally predatory animals, stalking and pouncing aggressively on their prey. This natural instinct can have painful results when a kitten sinks her needle-sharp teeth and claws into your ankles as you walk past her.

Case History

Joe brought home a cute little black and white kitten to surprise his eight-year-old daughter, Susie. Susie was thrilled and named the kitten Josie, after her father. All day she played with the kitten, and when she went to bed, she took Josie into her room with her. When Joe looked in on Susie before he went to bed, Josie was curled up at the little girl's feet. But the next morning, he was startled and alarmed to hear Susie shrieking, "No, Josie. No!"

When he rushed into her room, he found Susie crying, her ankles and legs covered with scratches, her pajama legs spotted with blood. "Josie jumped out from under the bed and grabbed my legs and scratched me," she sobbed. "I guess she doesn't like me."

During the day, the kitten played perfectly normally with Susie, but the same scene was repeated the next morning. Joe told Susie that they'd better get rid of the kitten, but she didn't want to, so he called me for advice.

I explained to him that the kitten didn't attack Susie because she didn't like her, but out of purely instinctive predatory behavior. When the child put her legs down out of bed, with her pajama legs flapping, it seemed to the kitten that this was some kind of animal waiting to be pounced on.

I made the following suggestion. To cure this unpleasant and painful behavior, Joe and Susie needed to let the kitten know right away that it was not allowed. I told Joe to give Susie a spray bottle filled with water. As soon as she began to put her feet on the floor and Josie leaped out from under the bed, Susie should immediately spritz her, in the face if possible, clap her hands, and shout "NO." Alternately, she could shake a can filled with pebbles or pennies and say

"NO." If Susie could do this every time Josie jumped out at her, the kitten should soon get the message that the behavior is unacceptable.

The element of surprise may also be at work here. If something that seems to be potential prey suddenly appears to a kitten with no warning, she will attack. So, another thing that may help Susie is to give her kitten some warning that she is about to put her feet down. She can loudly say, "Hi, Josie, I'm about to get up," or anything else that lets the kitten know it is her feet about to appear, not some potential prey. The kitten will probably change her predatory behavior to greeting behavior.

Redirected Aggression

This is another behavior that is often puzzling to cat owners. It is a reaction to the frustration of seeing an enemy nearby and not being able to do anything about it. The cat will then redirect her anger and attack the nearest target.

For example: Your cat, Sam, is lying on the windowsill, calmly looking out the window. Suddenly, he stiffens, growls softly, jumps off the windowsill, and viciously attacks the other cat in the household; or he may attack you, whichever is closest.

This is what has happened: He has seen a strange cat (probably another male) in his backyard, and because he cannot reach the other cat to chase him away, Sam has redirected his aggressive feelings to strike out at the nearest living creature. This behavior is closely related to *territorial aggression*.

If you happen to be in the room when your cat suddenly stiffens at the sight of a strange cat outside, immediately grab him and take him away from the window. If you cannot catch him in time and he attacks another household cat, simply separate them by closing one in another room until the aggressor calms down.

If this happens frequently, you may want to consider getting a shade, blind, or curtains to block the view from the window.

Territorial Aggression

Cats are naturally territorial. In the wild, female cats closely guard the area where they live and raise their kittens; male cats usually have much larger territories covering those of several females, and they will fight to protect their turf from other males. Neutering a male cat will greatly cut back on his territorial aggression.

But as in the case of Ike and Mike, even neutered cats retain a sense of territory and may fight with other cats (or indulge in marking behavior) if a new cat is introduced into a household, especially another male.

The solution to this behavior is to isolate the cats. Put the newcomer into a carrying case or crate so that the resident cat(s) can become used to her sight and scent. Have each cat, newcomer and longtime residents alike, take turns being confined in the same case or crate so that each cat becomes accustomed to the other's scent. Gradually let the newcomer out for short, supervised visits. Be sure to do this in an enclosed area—such as a room with the door closed—so that neither cat can run off and hide. The dominant, territorial cat will probably accept the newcomer eventually, but he may always bully her a bit. If this becomes too violent, you will have to step in and again isolate them.

Most territorially aggressive cats will learn to allow a feline newcomer to live at peace in a household, but in some cases it just will not work out and the newcomer will have to be relocated.

A strange and often troubling form of territorial aggression occurs when one cat in a household has been hospitalized and returns home to convalesce. Often, the cat that has remained home will attack her friend, seeming not to recognize her. This behavior is very upsetting and confusing, not only to the returning cat but to her owners as well. The problem is that cats recognize each other primarily by scent. The hospitalized cat has a totally unfamiliar smell—she smells of disinfectant, medication, and all kinds of foreign aromas—and her former friend simply does not recognize her and perceives her as a stranger who has invaded her territory. The solution again is to isolate the returning cat, in this case in a room with a closed door and all of the amenities (litter pan, food and water, sleeping place, etc.) and allow her to recuperate in peace. Again, mixing scents by swapping out the cats in the isolated room will speed up the process of acceptance. When the returning cat gets stronger and her natural scent returns, it is safe to allow her out into the house again. Her former friend will now recognize her and peace will reign.

Another troublesome form of territorial aggression is when a cat takes a dislike to a particular human, or to every other human except her owner. This may be caused by stress—in other words, the cat was never exposed to a number of different people as a kitten—but whatever the cause, it can be very disturbing when your pet cat attacks anyone else who walks into your home.

Case History

Richard was a bachelor and loved to entertain. But he had a problem with his beloved Rex cat, Chessie. As soon as friends arrived and the doorbell began to ring, Chessie would become tense.

At first, she simply retreated to the bedroom and sulked until Richard's guests left, but lately she had started to appear in the roomful of people. She would growl and hurl herself at the nearest person, sinking her teeth or claws into the ankles or legs of the startled guest.

Richard was appalled and confused. What should he do?

I suggested he get a large, comfortable crate and put it in the bedroom. Before his guests were due to arrive, he should place Chessie in the crate with a lot of reassurance and a treat or two. If he could arrange it so that his guests arrived at the apartment a few at a time instead of in one large, noisy group, it would be less stressful for the cat. After everyone had arrived, he could lift her out of the crate, hold her in his arms, and introduce her to a friendly guest or two. She should then go back into the security of the crate. This routine can be repeated as often as it takes for her to accept the presence of strangers in "her" home, but it is usually a good idea to have the cat in the crate during the confusion of guests' arrival.

Maternal Aggression

This form of aggression is one that an owner who plans to breed his pet should be aware of. A mother cat will always fiercely and aggressively protect her kittens from any perceived danger.

This means if you have a female with kittens, you need to shield her from any threats or she may strike out and attack to protect her kittens. Noisy, boisterous children and other household pets may seem to pose a danger. You should see to it that the mother cat and her kittens have a quiet, secure place away from unnecessary disturbances. Visits by curious and interested children and other pets should be kept to a minimum and always under your supervision. Visitors should be advised not to try to touch or pick up a kitten.

The kittens will grow up fast and soon leave the security of their mother's side. Then it is safe to touch and play with them gently without fear of maternal aggression.

Fearful Aggression

By the same token, a cat that is fearful for her own safety will strike out aggressively to protect herself. If a dog frightens a cat by badgering or cornering her, even if the dog is well meaning, she will scratch the dog in fear and self-defense. She also may scratch a child who teases, taunts, or grabs her. This is not "bad" behavior; it is fear behavior.

A thoughtful owner will try to protect a new cat or kitten from sudden, frightening confrontations until the cat has become accustomed to her surroundings and the other creatures there and is no longer afraid. Children and dogs need to be taught to treat the cat with respect and gentleness.

Pain-Induced or Medically Induced Aggression

A cat in pain will not want to be touched, handled, or even approached. One that is simply not feeling well due to a medical condition may also want to be left alone. If a formerly calm and friendly cat suddenly changes her demeanor and strikes out whenever you approach her, you should suspect a medical problem.

Case History

Peter was a cat that could only be described as a love. A big, rangy Maine coon, he was gentle, mellow, and affectionate. Usually, he calmly allowed "his" children to pet him, haul him around, and in general treat him like a stuffed animal. Suddenly, one day, he hissed and growled when the children approached him as he lay on the sofa.

Alarmed, his owner, Sofie, called me. What did I think was going on?

When I arrived at the house, Peter was curled up, asleep, on the sofa. When I approached him and put out a hand to stroke him, he hissed and growled softly, and seemed to shrink away from me. This was definitely not the gentle Peter I knew from previous visits. I sensed something was wrong and told Sofie to take him to the veterinarian.

Sure enough, Peter was suffering from a painful mouth abscess and was acting aggressively to protect himself from further pain. As soon as the abscess was cleared up, he returned to his usual mellow, affectionate self, willing to be picked up and hauled around by the children.

Cats and Babies—Preventing Aggressive Behavior

Cat owners who are first-time expectant parents are often concerned about how their pet will accept the new baby. Will she resent the baby and be aggressive toward it; might she harm the baby? People in this situation often consult me, and my advice is this:

If a cat has been an "only child," lavished with love and attention, it is a good idea to hold back a little before the baby's birth. In other words, do not nurture the cat quite so much—remember, you will have less time for her after the baby's birth.

After the birth, try to continue to have some quality time with the cat despite your hectic schedule. Encourage her to sit by your side and stroke her while you are nursing or feeding the baby. Allow the cat to smell the used diapers when you are changing the baby so the cat will realize he is a human and is part of you. Clothing that has been worn by the baby can be used as well, in the cat's bed. But diapers, because they have a strong smell, are necessary for the cat to make the connection.

Some owners put a screen door on the baby's room to prevent a cat from getting into the baby's bassinet or crib at night.

Go easy and do not force it, and usually a cat will quickly learn to accept the new baby graciously. Most cats get along very well with babies. But remember, the cat was there first and needs your protection as the child gets older and begins to grab. Even very young children can learn to be gentle with animals.

PLANT CHEWING OR DIGGING AND ELIMINATING IN PLANT POTS

The first thing houseplant lovers with cats should do is to be sure the plants they have around are not potentially harmful (discussed in Chapter Five). Cat owners who also have houseplants are often faced with the problem of ragged, chewed plant leaves. Worse, they may be horrified to discover that their cat is using the earth in their large potted plants as a latrine.

There are several things you can do to prevent a cat from chewing on your prized orchids. Many cats love to chew on greens. First, you can provide your pet with her own acceptable plant to chew. Plant a small pot with grass, alfalfa, or wheat seeds and place it on a windowsill. Most cats will really enjoy chewing on the tender sprouts as they come up and will appreciate being praised and stroked as they do so.

This alone may not prevent your pet from chewing on your favorite houseplants. If she persists, you will need a deterrent. The smell of mothballs placed in the earth around a plant will repel most cats. The trouble is, it may also repel you. There are bitter repellents on the market designed to spray on plant leaves and discourage animals from chewing them. They are harmless to plants but taste horrid to cats.

When a cat uses the earth in a large planter as a litter box, it not only creates an aesthetic and odor problem, it will eventually kill the plant. Here is the case of Maggie and her owner, Dorothy.

Case History

Dorothy was an avid houseplant gardener. She had beautiful plants all over her house. Her prized possessions were two huge flowering orange trees that had been her father's before he died. They stood in large clay planters on either side of her enclosed sun porch. She fed and nurtured them regularly, and they rewarded her with aromatic blossoms and wonderful oranges every year. Her other passion was her petite gray cat, Maggie. Maggie was the picture of ladylike, fastidious feline grace and charm.

Then one day Dorothy noticed that one of her orange trees had developed yellow leaves. Upon investigating, she saw that the earth around the tree had been disturbed. When she looked more closely, she was horrified to discover a piece of cat feces buried in the earth. Her refined little Maggie had obviously used the planter as a litter box. Then Dorothy remembered having found some earth around the planter a couple of times. She had attributed it to her own carelessness, but now she wondered—how long had this behavior been going on, and how could she prevent it without catching Maggie in the act? After giving the plant a lot of water to hopefully flush the toxins out of the soil, Dorothy called me.

Because she could not always be around to stop Maggie with a spritz of water or other corrective tactics, I suggested that she take a defensive approach to shock Maggie away from the planters. Mothballs, as I said above, are excellent repellents, but many people do not like their strong smell. Crinkled-up aluminum foil placed on top of the earth will usually send a cat elsewhere. But I think the very best repellent is a sticky surface. In the case of large planters, such as Dorothy's, I suggested she buy some Con-Tact paper, cut it to fit around the tree trunks in the planters, and place it sticky side up in the planters. The minute Maggie's paws hit the sticky surface, she would be surprised and repelled.

A week later Dorothy reported that all was well. Maggie hadn't gone into either planter, and the trees were now recovering.

DUMPING GARBAGE

This really need not be a problem if you secure your garbage in a lidded can and take the bag out whenever it contains anything that might appeal to your cat. But some cats are truly "garbage hounds," and some households are not geared for regular garbage removal. A cat that dumps over the garbage can is not being bad. She is just hungry, or maybe a bit mischievous and curious.

The best way to modify this behavior is to catch the cat in the act and seriously discourage her with a verbal "NO," hand clapping, a shake can, or a spritz of water.

The problem is that this behavior usually occurs when garbage is left overnight, or during the day when no one is around. If the garbage contains poultry, fish, or meat bones or scraps and a cat is left alone to her own devices, it becomes fair game. The most serious consequence of this activity, besides the mess it creates, is that bones of any kind can be extremely dangerous if not fatal for a cat to swallow. String that has been tied around meats or poultry and retains the aroma is another serious, possibly fatal, danger.

If you are unable to secure your garbage, the next-best solution is to set up a booby trap. Put mothballs in the garbage, or a few drops of ammonia or Clorox, or something in the top of the bag that will startle the cat when she puts her paws in the bag, such as a couple of blown-up balloons. Usually one or two experiences with something unpleasant and startling in the garbage bag will condition a cat to leave the garbage alone.

WALKING OR SITTING ON PROHIBITED SURFACES

Most cat owners recognize that cats love to explore every nook and cranny of their surroundings. They also know that cats are very agile and can go almost anywhere in a house that they want to. With this in mind, the majority of cat owners do not try to keep their pets off any surfaces in the house.

There may be exceptions, however. The beautiful red velvet sofa may be a no-no, or the étagère filled with delicate antique figurines. Most owners, understandably, do not want a cat up on a kitchen counter when they are preparing food, but this is usually not a problem because the owner is right there and can prevent the cat from jumping up. But some people do not want a cat on the kitchen counters at any time.

Sometimes owners mistakenly give a cat mixed messages about where it is all right to go. For instance, if a cat is regularly fed on a kitchen counter (usually because there's a dog in the house who would eat her food if it were on the floor), how is she to know that she is not supposed to get up there when her owner is cooking? If she regularly sleeps on her owner's bed, how can she tell the difference between the bed and the velvet sofa?

An owner should try to understand this confusion and realize the cat is not being "bad" when she walks or sits where she should not.

If you are around when the cat goes somewhere you don't want her to, a spritz from a water bottle, a firm "NO" accompanied by a hand clap or a rattle from a shake can—anything that will startle her and make her stop whatever she is doing—will eventually condition her to stay away from the forbidden surfaces.

But if you are away from home a lot, you will have to deter her in other ways.

Case History

Eleanor lived alone and was a fastidious housekeeper who worked in an office every day. She had never had a pet but decided she would like a kitten. After much searching, she finally found the perfect little white kitten. She named him Wilbur. She brought him home on a Friday night so they would have the weekend together to become acquainted.

Everything went wonderfully. He immediately used the litter pan, ate his dinner, and slept on Eleanor's bed. She was delighted.

The trouble started when she came home from work on Monday. To her horror, there were little paw prints on her gleaming kitchen counters. Wilbur had been up on them! She was horrified and immediately called a cat-owning friend to ask what to do. The friend was slightly amused but realized that Eleanor really could not live with a cat that walked on kitchen counters, so she called me for advice.

I suggested she tell Eleanor to buy a big roll of Con-Tact paper, cut it to fit the countertops, and put it sticky side up on the counters. It worked like a charm, and Wilbur and Eleanor are still happily sharing their home with no paw prints on the kitchen counters.

Alternately, crinkled-up aluminum foil placed on the countertops may work—but some cats do not mind foil, and many simply play with it and push it off onto the floor.

KEEP IT SIMPLE

As you can tell by many of the solutions I have offered for cat behavior problems, the simpler the solution, the better. Most behavior problems, if they are not prevented to start with, can be solved with a little common sense on the owner's part. If an owner can learn to understand why a cat is behaving in a certain way, he can usually figure out how to modify the cat's behavior with a lot of Patience, Persistence, and Praise and a good helping of CatSpeak.

Chapter Ten

On to the Future: More Things You and Your Cat Can Do Together

You may be content to have your cat simply be a loving pet that never leaves home and keeps you company while looking decorative. Your cat may also be content to just hang out in the house, sleep a lot, and watch birds and squirrels from the window. But if you and your cat enjoyed the simple training steps I outlined in Chapter Eight, you may want to widen your horizons.

Many outgoing, inquisitive, and active cats naturally enjoy the stimulation and challenge of going to new places, meeting new people, and learning more skills, just as dogs do. Some may need to learn to enjoy it. Other cats are shy, a bit fearful, or quiet and really do not want to leave the safety of their homes. By now you should know your own cat's personality and be able to judge whether or not she is willing and able to enjoy new experiences.

In this chapter, I will explore various activities an outgoing cat may enjoy. They go a step beyond normal training—on to the future! The subject I'll talk about first is travel with your cat. Second, I will tell you how to help your cat become a pet therapist. Third, we'll talk about participating in cat shows. And fourth, I'll show you some more advanced tricks you can teach your cat for enjoyment. She might even eventually become a "star"—a cat model for print advertisements, television, or movies.

TRAVEL WITH A CAT

The myth that cats are devoted to only one location and are not happy away from it does not apply to all cats. I know owners who take their cats everywhere with them—to hotels, resorts, or visiting in other people's homes. They go on subways, trains, planes, and in cars perfectly calmly and quietly. Other cat owners regularly take their pets to stay in summer cottages or winter ski lodges. These cats enjoy being in new places with their owners, rather than being left home alone. Of course, you will need to take along all of your cat's supplies: food and water, bowls, toys, a litter pan, a small scratching post, and so forth.

A cat resting comfortably in her carrying case.

If your cat will travel, she must have a sturdy carrying case. To make the process easy for both of you, you will have to teach her to hop right into the case whenever you ask her to. She also should have learned to respond to her name. For her safety and security, she needs to wear a harness or collar and leash. (I talk in detail about all of these things in Chapter Eight.)

Before you take off on an extended trip with your cat, you will have to practice. Put her into the case and take her to a friend's or family member's home. Let her out of the case and allow her to explore. If she seems nervous in new surroundings, call her, give her a treat or two, and pet and praise her. When it is time to go home, call her back into her case. When you get home and she jumps out of her case, praise her again and give her another treat. Do this as many times as necessary in different locations until she calmly accepts whatever new surroundings she finds herself in.

If you are going to travel by car, take her on frequent, short car trips so she gets used to the motion, noise of traffic, and smell of gasoline fumes. Make each trip a happy experience, with a treat and playtime or other reward when you return home.

The same is true if you will be going by train. Go on short train rides, perhaps to the next station and back again, so she becomes accustomed to the different motion of a train and all of the sounds and smells surrounding train travel. It is more difficult to practice an airplane ride, but trips in a car are somewhat similar.

When your cat is perfectly happy to jump into her carrier and take off with you, you are ready for a real trip. I suggest you keep the first trip short. Perhaps you can spend one night in a local hotel or motel, or a weekend in a rented cottage.

As soon as you are sure your cat will remain calm and responsive to you while you travel and stay in an unfamiliar location, you will be able to enjoy her company (and she yours) wherever you go.

PET THERAPY

Well-behaved and trained cats are wonderful "therapists" in nursing homes, children's hospitals, and senior residences. They are soft, pliant, and quiet and are welcome equally on a lap or a bed.

For a cat to become a good therapist, she must learn to go willingly into her carrying case and to travel with ease. She should also have learned to wear a harness or collar and leash. This is required in most care facilities to ensure that the owner or handler is always in complete control of the cat.

In Chapter Eight, I showed you how to teach your cat to lie down and stay in a lying position. It is very important for a therapy cat to learn to do this so that she will remain quietly on a blanket or pillow on a patient's lap or on a bed until her owner/handler lifts her off, no matter what the distractions might be. These can include strange wheeled objects such as wheelchairs, IV stands, and walkers. She will also need to stay calm if a sudden action or noise occurs, such as something dropping to the floor, or someone coughing explosively or talking very loud. You will need to work with her to accustom her to these things. Some cats are not fazed by anything, but others will require quite a bit of conditioning to become used to the people and noises they will encounter in a care facility.

The only way to condition a cat to these things is by acting out the situations that might occur and reassuring the cat until she is not disturbed. Borrow or rent a wheelchair and other wheeled objects. First have the cat, wearing her harness or collar, sit on a cushion on the lap of a friend seated in the wheelchair. Say "Stay" and have the person move around, backward, forward, left and right, while stroking and reassuring the cat. You should stand nearby and constantly reassure the cat. Once the cat relaxes and stays still and quiet during the wheelchair's motions, praise her and stroke her. She has now learned to stay quiet and calm in a moving wheelchair.

Next, with the cat still on someone's lap, bring a wheeled cart or walker up to her while reassuring her. Some cats pay little attention to the approach of a wheeled object; others may stiffen and want to bolt (flight behavior). Reassurance in the form of verbal encouragement and petting will usually calm a cat. Practice this movement as many times as necessary until the cat remains quiet and calm whenever any wheeled furniture approaches her. It may take a while to condition a cat to these foreign objects, but it is absolutely necessary if she is going to participate in therapy work.

Some cats startle easily—the slightest unexpected sharp noise may cause them to stiffen and flee. A therapy cat cannot react this way. She must remain calm and unruffled if a book, metal tray, spoon, or cup suddenly crashes to the tile floor (not an uncommon occurrence in nursing homes). Practice over and over with your cat—drop spoons, knives, or tin trays on the floor and immediately reassure her with your voice: "It's all right," "Don't worry about it," "Stay," "Be quiet," and so forth. Reach down and stroke her. Reward her with a treat when she remains relaxed. With some cats, this may take a great deal of practice.

Many care facilities routinely ask residents to sit in a circle and pass objects around. When a cat is visiting, she may be passed around the circle from lap to lap. It is necessary for your cat to accept this type of activity. Invite family and friends to sit in a circle with pillows or blankets in their laps. Pass the cat around the circle. She should remain calm and relaxed. Have each person pet and reassure her until she begins to perceive this as an enjoyable experience.

Once your cat learns to accept all of these possible care-facility experiences, she will be well on the way to becoming a qualified therapy cat. The first few times she visits a care facility, it is extremely important for you to remain very close to her so you can reassure and praise her constantly and correct her if necessary. When she begins to be comfortable in her new role and revels in it, you will eventually be able to stand across the room and watch her "do her thing."

CAT SHOWS

No matter what kind of cat you have, random-bred or pedigreed, your pet can participate in cat shows. Cat shows have been modernized and now allow all types of cats to enter in various categories. Even if you don't have a pedigreed show cat, you may think your cat will enjoy showing off. If you also think her coat and body type are wonderful-looking, you might want to enter her in a cat show.

Cat shows are crowded, not only with all kinds of cats but with all kinds of people. They tend to be noisy and hectic. A cat that is very shy and wary of strangers will not enjoy or do well in a cat show. But a social, outgoing cat may thrive in this environment.

There are several things you can do to prepare your cat for a cat-show experience. Again, she must be used to traveling in her carrying case. She should be accustomed to being touched and handled by strangers. This is something you can practice at home. With your cat in a sit-stay or down-stay position on a tabletop, invite an acquaintance to stroke her and examine her body, inside her ears, under her tail, and so forth. Have him pick her up and hold her in the air. If she resists these attentions, step in and reassure and pet her. Do this over and over again, with different people acting the part of a judge if possible, until your cat is perfectly comfortable being handled by a stranger.

She also has to learn to relax while you carry her through a crowded room over to a judges' table. There is a prescribed way to carry a cat in a show: Support her front legs with one hand and her back legs with the other while you stretch her body horizontally to its fullest length. Practice this again and again until your cat is able to remain quiet while you carry her in this way.

Once your cat has learned to enjoy being handled by strangers and to accept being carried through a crowded room, she is ready to enter a show. For information about where and how to enter your cat in a show, go to the Cat Fanciers' Association website (www.cfainc.org) or log on to my website at www.pawsacrossamerica.com.

MORE TRICKS

In Chapter Eight, I taught you a few simple cat tricks. If you and your cat enjoyed learning these tricks, you may want to go on to learn some more difficult ones. By the way, tricks are wonderful for pet therapy. Children and seniors—in fact, most people—really love to see a cat perform tricks.

Most of these tricks should be taught with the cat on a tabletop. Training sessions should be short and always end on a happy note. I like to train my cats just before their dinnertime. That way, the training session ends on a very positive note. Alternatively, you can give your cat a treat after training. Always praise your cat after a session, even if it has not been especially successful.

Before you begin to teach your cat a new trick, always review all of the things she has learned previously, beginning with sit-stay and down-stay and any other tricks she knows from Chapter Eight.

Roll Over

This trick should *not* be learned on a tabletop, for obvious reasons. With the cat on a carpeted floor, put her in the down-stay position. Squat down and hold one hand across her paws so she cannot move. With the other hand, hold a treat over and just behind her head and move it around. When she turns her head to focus on the treat, let go of her paws and say "Roll over." She will naturally turn over to get the treat. When she rolls over, praise her and give her the treat. (Note: You can begin this trick with the cat on either side so she rolls from right to left or left to right.) Do this over and over until you no longer need to hold her paws and she rolls over on command. Then you can stand up and give the command.

Jumping Through a Hoop

Put the cat in a sit-stay position on a tabletop and place the hoop flat on the table in front of her. Put a treat in the center of the hoop. As the cat walks into the center of the hoop to get the food, say "Jump." Then gradually lift the top edge of the hoop as you say "Jump" and the cat walks through the hoop to get the treat.

Continue to lift the hoop higher until it is above the surface of the table. Now the cat will have to jump through it in order to get to the treat on the other side. Each time the cat jumps through the hoop, praise her. Eventually you may be able to go one step further and hold the hoop over a second table. The cat will jump from one table to another through the hoop.

There is another way to teach a cat to jump through a hoop. Instead of beginning with the hoop lying flat on the table, hold it upright and encourage her to walk through it as you offer a treat on the other side and say "Jump."

A variation of this trick is to have the cat jump onto your shoulder. To do this, place a treat on your shoulder, stand with your back to the cat, tap your shoulder, and say "Jump." The cat will jump from the tabletop up onto your shoulder.

Waving

This is an extension of shaking hands, which I described in Chapter Eight. With the cat sitting on a tabletop, have her shake hands and give her a treat. Then move back with a treat in your hand and say "Shake." The cat will extend her front leg out in an effort to reach the food. Then say "Wave" as you move farther back. When she continues to wave, give her the treat and praise her. Continue to do this until she learns to wave on command.

Retrieving

Some cats are very good retrievers; others never seem to get the hang of it. If you start when your kitten is young, she will learn this trick more quickly. Tie a favorite toy, feather, or piece of crumpled paper to the end of a long string. Toss the object a distance away. When the kitten pounces on it and takes it in her mouth, say "Fetch" and pull gently on the string. When the kitten is right in front of you, take the object out of her mouth, praise her, give her a treat, and toss it again. This time, say "Fetch" as she runs toward the object. Repeat the reeling-in and tossing-out motions again and again, each time praising the kitten when she lets go of the object. After awhile you may no longer need to pull on the string; your kitten will run right back to you with her "treasure" so she can have a treat and the fun of having you toss it again.

Be careful! A cat that really loves to fetch may constantly badger you to toss an object for her.

Speaking

Some cats have a natural ability to speak loudly and a lot—Siamese fall into this category. Others have barely any voice at all and simply open their mouths wide with no sound. My Persian, Chickie, had what I call a "silent" meow.

The best way to teach a cat to speak is to pay attention to the times when she meows naturally. Perhaps she always meows as you are opening her food. If she does, say "Speak" as you prepare her dinner so she will learn to associate the word with meowing. Maybe she meows when she wants you to groom her. Again, use the situation to teach her the word so that she will meow whenever you say "Speak."

SHOW BUSINESS

If your cat has learned a number of tricks easily, you may decide it's time for her to graduate to bigger things. You may want to see about getting her into show business—posing for print advertisements or appearing in television commercials. Just be sure your cat has the ability and willingness to do so. She must be the kind of cat that enjoys meeting new people, doesn't at all mind being handled by strangers, and is not fazed by noise and confusion.

She must also become accustomed to lights and cameras. Practice with the cat on a table while she is wearing a harness or collar and leash, and tell her to stay. Have someone repeatedly come by with a flash camera while you stand nearby and reassure the cat. Sometimes it helps a cat to focus on the camera if you or the photographer dangles a toy on a string in front of the lens. When the cat looks at the toy, pull it away and take the picture.

Some Additional "Tricks"

You can use your knowledge of typical cat behavior to manipulate your cat into certain actions and make it look as if she is doing them on command. This is a wonderful asset for a "show business" cat.

For instance, if you want your cat to walk with you, use the same technique you used to make her wave. Dangle a treat in your hand, just out of the cat's reach, and walk along. The cat will naturally walk beside you in order to follow the treat.

Maybe you would like her to reach out a front paw and "choose" something off a grocery store shelf, as I did for the Tender Vittles commercial. To do this, hold the cat in your arms and restrain all of her paws. As you approach the shelf, let go of one front paw. The cat will automatically reach out with her front paw in order to climb the shelf. You have used typical cat behavior to make it look as if you have taught your cat to grab food off the shelf.

These "tricks," and all of the other tricks you and your cat have learned, may make her into a media star—but *only* if she wants to be.

To find out how to start your cat on a media career, log on to my website: www.starpet.com.

CAT OLYMPICS?

I would really like to see some kind of Olympics for cats. These games would consist of agility demonstrations during which cats and their owners would have a good time.

I envision cats walking up ramps, going through tunnels, jumping through hoops, and climbing fabric-covered poles (cats with claws, that is). I feel this would take cats to a higher level of social experiences and create a whole new way of thinking about cats and their abilities.

Think about it!

Appendix: CFA Feline Good Citizen Award

The Cat Fanciers' Association (CFA) Feline Good Citizen (Feline Lifetime Partner/Companion) award/tests are a good way to judge your cat's well-mannered behavior. These tests can be given to either pedigreed or random-bred cats. The cats are asked to perform in everyday situations. (I have shown you how to prepare your cat for some of the tests—especially in Chapter Eight.)

The evaluator, or person who gives the test, should be someone who is not known to the cat. He should not be the owner, a member of the owner's family or household, or a regular caretaker. The tests should be given in a neutral spot outside of the cat's home territory. Once a cat has successfully completed the four basic tests, she will become a Feline Good Citizen and will be on her way to qualification for pet therapy and showing.

CFA FELINE GOOD CITIZEN TESTS

The Four Basic Tests

Test One: A Cat Accepts the Approach of a Friendly Stranger This test demonstrates that a cat can accept a friendly stranger, allowing him to approach and speak to her and her owner without showing aggression, panic, and/or flight behavior.

The cat is on a tabletop. The evaluator approaches the cat while the owner stands nearby. The cat remains calm and quiet. The owner can reassure the cat if necessary.

Test Two: A Cat Stays Quietly and Allows a Friendly Stranger to Pet Her Gently and Pick Her Up This test shows that a cat will allow a friendly stranger to pet her gently and pick her up while her owner stands nearby. The cat will remain calm and relaxed and will demonstrate neither aggression nor panic.

The cat again starts on a tabletop. The evaluator gently pets the cat, then picks her up and holds her loosely while the owner stays near. Again, the owner can reassure the cat if needed.

Test Three: A Cat's Appearance and Grooming This test shows that the cat's owner gives her good care and is concerned for her well-being. The cat appears well groomed and cared for. Her coat is shiny and neat, her eyes and ears clear, clean, and free of foreign matter. Her claws are well trimmed. The cat allows a stranger to perform minimum grooming tasks, touching her and then combing or brushing her lightly.

Test Four: A Cat Readily Goes into Her Carrying Case This test is very important, because a cat is often required to go into her carrying case. The owner stands across the room from the cat, next to an open carrying case on the floor. The owner encourages the cat to come. When the cat comes, the owner shows her the open case, and the cat should calmly go into it.

CONGRATULATIONS!

When your cat has passed these four basic tests, she is truly a Feline Good Citizen, and you both should be proud! She is well prepared to travel with you and stay in hotels or other people's homes comfortably and calmly.

MORE?

You may want to go further with your cat and prepare her to participate in pet-assisted therapy at nursing homes or other care facilities. You might also want to enter her in cat shows.

Earlier, I showed you how to prepare your pet for these more demanding roles; here are some additional tests for her at this stage: Tests five, six, and seven are primarily for therapy cats; test eight can also be useful for therapy and is necessary for a show cat.

Test Five: A Cat Allows Herself to Be Harnessed (or Collared) and Leashed and Carried into a Roomful of Strangers Without Showing Fear or Panic This test is primarily for a cat's safety in a strange place and also for manageability in a situation such as a nursing home or other care facility (it is required in many establishments). A cat that has been taught to wear a leash well can be carried into a group of people and remain calm and under her owner's care at all times.

Test Six: A Cat Remains Calm When Confronted by Distractions Such as Loud Noises, Sudden Movements, or Unfamiliar Objects Such as Wheelchairs This test is important if a cat will be a therapy pet in a nursing home or hospital and confronted with a number of previously unknown and strange objects. In this test, the cat wears a harness or collar and leash. She stays relaxed (her owner can reassure her) as someone approaches with an unfamiliar object such as a wheelchair, IV stand, or walker. She also stays calm if she hears a sudden, sharp noise such as a dropped book or other object. Continuous reinforcements in the form of petting and verbal encouragement are helpful.

Test Seven: A Cat Allows Herself to Be Placed on the Lap of a Gentle Stranger and Stays Quietly in Place Until Her Owner Removes Her This test is also necessary for a cat to perform pet therapy. Her owner will carry her to a seated stranger and place her on the stranger's lap on a blanket or pillow (to protect the frail person's body), stroking her from head to tail and reassuring her. When he places the cat on the person's lap, he says " Stay" and walks a short distance away. The cat remains quietly on the stranger's lap until her owner lifts her up.

Test Eight: A Cat Allows Herself to Be Carried Through a Crowded Room A show cat must pass this test. The owner holds the cat, supporting her with both hands: one beneath her front legs, the other beneath her back legs. The cat's body is stretched out horizontally in proper position. The cat remains calm and quiet while she is carried through a crowded room.

Glossary: A Few Terms Used Frequently in Reference to Cats

Brachycephalic: Having a round head, short pug nose, small nostrils, and prominent eyes (e.g., Himalayan).

Dewclaw: Extra claw found on inner front legs.

Estrous cycle: Heat cycle.

Estrus: Heat period.

Feral cat: A wild or untamed cat; in a household situation, a cat that is somewhat feral will be "scary" with strangers and is often dominant over other, more domesticated cats.

Nocturnal: Active at night.

Pica: Eating nonfood items, such as wool.

Polydactyly: Inherited trait resulting in extra toes, usually on front feet.

Queen: Female breeding (unspayed) cat.

Tomcat: Unneutered adult male cat.

Urine spraying: Act of spraying urine backward, against an upright object; most common in tomcats, but females and neutered males may spray, since territorial marking is often due to stress.

Vulva: External female genitalia.

meow meow meow meow

Index

A

B

C

D

E

F

G

H

N

O

P

Q

R

S

T

Bash Dibra's unique ability to communicate with animals has won for him a global reputation and established him as a trainer of record for the animal companions of some of the world's most high-profile celebrities, among them Jennifer Lopez, Matthew Broderick, Sarah Jessica Parker, Martin Scorsese, Mariah Carey, and Kim Basinger. A recipient of the New York State Humane Association Award, the New York City Veterinary Medical Associations Unsung Hero Award, and the North Shore Animal League America's Legendary Bond Award, Dibra dedicates countless hours to animal-related causes. He is a board member of New York SAVE, a nonprofit organization devoted to saving animals in veterinary emergencies, and consults for the ASPCA, Bide-A-Wee, and the Humane Society of New York. Dibra lives in Riverdale, New York, with six dogs, four cats, and a bird. You can contact him at www.pawsacrossamerica.com and at bash@pawsacrossamerica.com.

Elizabeth Randolph, a former pet care columnist for *Family Circle* magazine, is the author of more than a dozen volumes about pets and their care. Her feline books include *How to Be Your Cat's Best Friend, The Complete Book* *of Cat Health* (with the New York City Animal Medical Center), and *The Veterinarians' Guide to Your Cat's Symptoms*. She and Bash Dibra previously collaborated on *Dog Training by Bash* and *Teach Your Dog to Behave*. A member of the American Society of Journalists and Authors, the Authors Guild, the Dog Writers Association of America, and the Cat Writers' Association, she is a recipient of the Dog Writers' Maxwell Medallion. Randolph lives in suburban New York with a Norwich terrier and two British shorthairs.